Cases on Circular Economy in Practice

Pietro De Giovanni
Luiss University, Italy

A volume in the Advances in
Industrial Ecology (AIE) Book Series

IGI Global
PUBLISHER of TIMELY KNOWLEDGE

Published in the United States of America by
IGI Global
Business Science Reference (an imprint of IGI Global)
701 E. Chocolate Avenue
Hershey PA, USA 17033
Tel: 717-533-8845
Fax: 717-533-8661
E-mail: cust@igi-global.com
Web site: http://www.igi-global.com

Library of Congress Cataloging-in-Publication Data

Names: De Giovanni, Pietro, editor.
Title: Cases on circular economy in practice / Pietro De Giovanni, editor.
Description: Hershey, PA : Business Science Reference, [2023] | Includes bibliographical references and index. | Summary: "This book focuses on circular economy and provides an analysis on how circular economy systems are implemented and managed in practice, which strategic alliances firms pursue, and how they coordinate through negotiation and agreements"-- Provided by publisher.
Identifiers: LCCN 2022018286 (print) | LCCN 2022018287 (ebook) | ISBN 9781668450017 (hardcover) | ISBN 9781668450024 (paperback) | ISBN 9781668450031 (ebook)
Subjects: LCSH: Recycling (Waste, etc.) | Circular economy. | Strategic alliances (Business) | Negotiation.
Classification: LCC TD794.5 .C374 2023 (print) | LCC TD794.5 (ebook) | DDC 363.72/82--dc23/eng/20220622
LC record available at https://lccn.loc.gov/2022018286
LC ebook record available at https://lccn.loc.gov/2022018287

This book is published in the IGI Global book series Advances in Industrial Ecology (AIE) (ISSN: pending; eISSN: pending)

British Cataloguing in Publication Data
A Cataloguing in Publication record for this book is available from the British Library.

Advances in Industrial Ecology (AIE) Book Series

ISSN:pending
EISSN:pending

Editor-in-Chief: Syed Abdul Rehman Khan Xuzhou University of Technology, China

MISSION

The **Advances in Industrial Ecology (AIE) Book Series** examines current, state-of-the-art studies in the areas of Industrial ecology and Circular economy. Industrial ecology is a rapidly growing field that systematically examines local, regional, and global materials along with energy uses and flows in products, processes, industrial sectors and economies. It focuses on the potential role of industry in reducing environmental burdens throughout the product life cycle from the extraction of raw materials, to the production of goods, to the use of those goods and to the management of the resulting wastes. Industrial ecology is ecological in that it (1) places human activity -- industry in the very broadest sense -- in the larger context of the biophysical environment from which we obtain resources and into which we place our wastes, and (2) looks to the natural world for models of highly efficient use of resources, energy, and byproducts. By selectively applying these models, the environmental performance of industry can be improved. Industrial ecology sees corporate entities as key players in the protection of the environment, particularly where technological innovation is an avenue for environmental improvement. As repositories of technological expertise in our society, corporations provide crucial leverage in attacking environmental problems through product and process design.

The **Advances in Industrial Ecology (AIE) Book Series** encompasses empirical, analytical, theoretical, reviews, communications, and conceptual research, and comprehensive reviews of relevant research case studies of effective applications in the areas of Industrial ecology and circular economy. It provides a practical and comprehensive forum for exchanging scientific knowledge and ideas which bridge the technological advancement and sustainable practices with industrial ecological management. The goal of this series is to bring together the expertise of professionals who are working in the field of industrial ecology and sustainability across the globe.

COVERAGE

- Resource Management
- Industry 0.4 and Sustainability
- Energy
- Recycling
- Circular Economy
- Sustainability
- Technological Advancement in Industrial Ecology
- Green Practices
- Material Efficiency
- Urban Metabolism

IGI Global is currently accepting manuscripts for publication within this series. To submit a proposal for a volume in this series, please contact our Acquisition Editors at Acquisitions@igi-global.com or visit: http://www.igi-global.com/publish/.

The Advances in Industrial Ecology (AIE) Book Series (ISSN pending) is published by IGI Global, 701 E. Chocolate Avenue, Hershey, PA 17033-1240, USA, www.igi-global.com. This series is composed of titles available for purchase individually; each title is edited to be contextually exclusive from any other title within the series. For pricing and ordering information please visit http://www.igi-global.com/book-series/advances-industrial-ecology/270446. Postmaster: Send all address changes to above address.

Titles in this Series

For a list of additional titles in this series, please visit:
http://www.igi-global.com/book-series/advances-industrial-ecology/270446

Green Chemistry for the Development of Eco-Friendly Products
Kavita Shakya Chahal (Government College, Bichhua, India) and Twinkle Solanki (Government College, Bichhua, India)
Engineering Science Reference • © 2022 • 276pp • H/C (ISBN: 9781799898511) • US $250.00

Innovative Nanocomposites for the Remediation and Decontamination of Wastewater
Azad Kumar (M.L.K. P.G. College, India)
Engineering Science Reference • © 2022 • 335pp • H/C (ISBN: 9781668445532) • US $240.00

Handbook of Research on SDGs for Economic Development, Social Development, and Environmental Protection
Cristina Raluca Gh. Popescu (University of Bucharest, Romania & The Bucharest University of Economic Studies, Romania)
Engineering Science Reference • © 2022 • 531pp • H/C (ISBN: 9781668451137) • US $325.00

Integrating Blockchain Technology Into the Circular Economy
Syed Abdul Rehman Khan (Xuzhou University of Technology, China)
Business Science Reference • © 2022 • 307pp • H/C (ISBN: 9781799876427) • US $250.00

For an entire list of titles in this series, please visit:
http://www.igi-global.com/book-series/advances-industrial-ecology/270446

IGI Global
PUBLISHER of TIMELY KNOWLEDGE

701 East Chocolate Avenue, Hershey, PA 17033, USA
Tel: 717-533-8845 x100 • Fax: 717-533-8661
E-Mail: cust@igi-global.com • www.igi-global.com

Table of Contents

Detailed Table of Contents

How circular economy systems truly work for firms around the world is at the beginning of knowledge development. As such, this chapter aims to provide an analysis of how to concretely implement and manage innovative projects to shift from a linear to circular supply chain management. The chapter analyses the case of sustainable wooden packaging logistics implemented by Fercam Echo Labs, moving from recycling approaches to upcycling solutions for pallets and crates. The development of circular supply chain management inside the circular economy can properly guide research and practitioners' efforts in the innovative logistics packaging management arena.

Lithion Power is India's leading provider of battery as a service, supplying lithium-ion batteries for e-bikes and three-wheelers. They use the battery swapping technology to power electric vehicles. Lithion's omni-charging technology allows electric vehicles to either charge at home, at public chargers or at a Lithion Swapping Point (LSP), primarily spread across North Delhi and the nearby state of Haryana. At these LSPs, a customer can drive in with their existing battery and in a less-than-five-minute walk away with a charged battery. Lithion is collaborating with major

quality social standards. Given the complexity of the business model, Fairphone has created a unique network of stakeholders, with customers playing a pivotal role, each providing fundamental help. Given the innovative and holistic approach Fairphone is implementing, the company could also generate positive effects in the growing field of smart cities. The chapter is dedicated to the analysis of Fairphone's business case and the positive impacts the company could generate within smart cities.

The chapter aims to gather information and analyze the new business model developed by Levi Strauss and Co. First, it analyses the network that LS&Co. has set up and how the business works. Then, it will investigate the context in which LS&Co. operates by looking at values, drivers, stakeholders, barriers. Subsequently, a careful analysis of the performance indicators that LS&Co. has adopted including the GRIs material report will be conducted. The technologies and developments that LS&Co. has adopted in its business will be explained; particular attention will be given to emerging partners in the new business model and the value they bring. There will be space for an ambitious project based on blockchain technology as a tangible tool for their products and suppliers. Some recommendations on how LS&Co. could improve its circular economy in the context of the growing smart cities will be given in the last paragraph.

Sustainability is an extremely urgent and universal concern; however, becoming sustainable doesn't happen overnight. It is not an immediate transition, and in order to really take sustainability to the next level, experience and know-how must be

Sustainability is significantly important to the fashion business because of consumers' growing awareness of the environment. When a fashion company aims to promote sustainability, the main link is to develop a sustainable circular system. This chapter contributes understanding of how a historical fashion company has managed to evolve over time by implementing circular technology that can give it a competitive advantage in the market. The authors firstly describe the structure of H&M, the value it distributes in the market, and the future goals it has set for itself. Next, they introduce the methodology by which the company has managed to make its circular model impact on its performance, giving an overview of the relationship the company has with its stakeholders and consumers. Furthermore, based on secondary data and analysis, they learn how the Swedish fast fashion company has built its sustainable strength by developing eco-friendly materials, monitoring sustainable production, reducing carbon emissions in distribution, and promoting circular fashion.

This chapter elaborates on the application of circular economy systems in the textile industry, which is historically influenced by seasonality effects that induce a search for circular solutions independent of environmental constraints and regulations. This chapter shows that circular economy systems can create business opportunities as well as important cannibalization effects when consumers are confronted with new and used textile goods. Hence, it highlights how the use of digital technologies can help to support the mitigation of such issues..

Preface

The current wave of the circular economy (CE) has become very popular in recent economies, since it is a valid driver in facing environmental and social issues, as well as a strategy involving multiple stakeholders – including people, governments, and firms. CE represents the most effective option for mitigating the negative effects linked to traditional environmental concerns (e.g., global warming, emissions, pollution), as well as more recent and challenging questions, such as the worldwide shortage of raw materials and energy, the pandemic, and international conflicts. These factors enhance the urgency of implementing CE systems, knowing that their effectiveness and success are strongly related to whomever remains on the bottom line and/on the last tier of the supply chain, such as consumers. In fact, CE can properly function only by receiving inputs from the entity possessing the goods and the objects to inject within the reverse system (De Giovanni & Zaccour, 2022).

In this framework, De Giovanni and Folgiero (2022) have developed the Waste Cascade Model, which is the evolution of the waste pyramid proposed by the Ellen MacArthur Foundation (2015). The cascade, which is displayed in Figure 1, finds roots in the tripartite "reduction, reuse, and recycle" concepts, which are key ingredients of the new strategic paradigm, along with value. The latter represents the key strategic lever to engage stakeholders in implementing and managing a CE system: when the value of returns is low, the economic convenience of closing the loop decreases accordingly, making CE less attractive. Hence, the role of all stakeholders, including governments, is vital in aligning the tangible (e.g., appropriateness of the returns' residual value) and the intangible (e.g., CE enhances brand value) motivations. When all motivations are aligned and the incentive mechanisms are properly set, the system moves from the linear to the circular.

The Circular Cascade Model identifies traditional circular production and business models by acquiring and managing resources that are considered waste on some levels of the supply chain and finding new paths for the creation of value, also including social and environmental spheres. The model suggests the set of actions that firms and stakeholders can evaluate to detect the best sustainable and circular strategy according to the stages of waste management, which are exemplified by sustainability targets, product value, usage opportunities, and product lifecycle. Therefore, the Waste Cascade Model drives toward the identification of a sustainable strategy by obtaining the maximum value from products, while reducing the amount of waste in the lifecycle and optimizing the circular economy practices through the analysis of possible usages. Within the cascade, the traditional options given by "reduction, reuse, and recycle," which comprise the circular economy systems, are complemented by additional business and strategic considerations.

Hence, the circular economy literature (see Ghisellini et al., 2016) has developed the tripartite "reduction, reuse, and recycle" paradigm as follows:

Reduction: Efficiency pursued through the minimization of inputs,
Reuse: Usage of parts and components linked to the closed loop flow,
Recycling: Recovery of materials from waste and their injection into production processes.

While in waste management the options "reduction, reuse, and recycling" are a vital part of the waste management pyramid, the Waste Cascade Model offers new and comprehensive ways to reengineer the circular economy strategy by providing a framework to easily identify the best circular option. This includes both the returns' residual value as well as the life cycle stage. The best CE option to implement should, in fact, definitely consider these two dimensions. In principle, returns' residual value and product life cycle move together over time. Therefore, when a good is returned very early (late) in its life cycle, the chance to receive a high (low) residual value are high as well, along with the possible activation of reusing (recycling) options. However, when the goods are returned in the early stage, but their residual value is low, the reuse option may be not feasible; for example, returning a new vehicle after an accident may require recycling rather than reusing. Similarly, when goods are returned in the last stage of their life but their residual value is still high, the recycling option may be not feasible; for example, returning an old vehicle that is still in great conditions can activate the reuse options through the secondary market rather than the recycling. Therefore, according the Waste Cascade Model by De Giovanni and Folgiero (2022), the CE paradigm to be selected depends on the strategic levers behind the whole business model.

Figure 1. The waste cascade

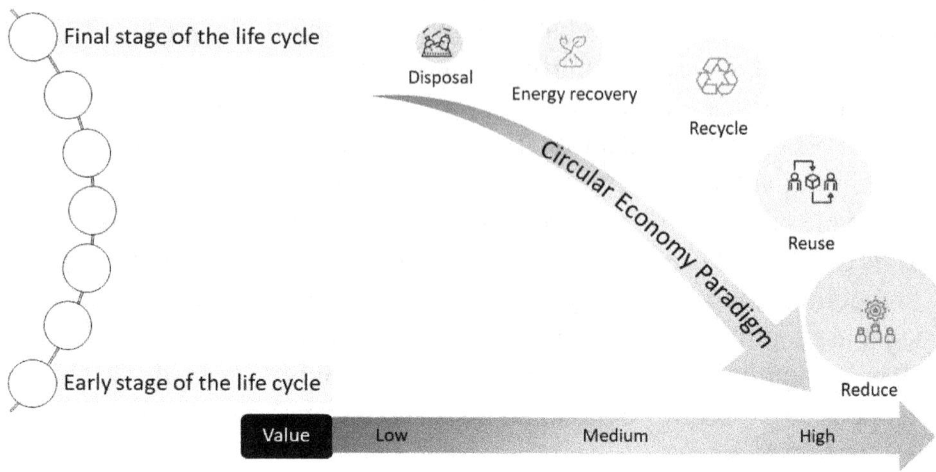

CIRCULAR ECONOMY IN PRACTICE

Although theoretical models and frameworks have been widely proposed in the scientific community, the enthusiasm and the interest surrounding CE have also been followed by skepticism and uncertainty due to the connection that the system should have with many parties and stakeholders, including consumers, as well as the stochasticity linked to the quantity that will be returned, the quality and the residual value of CE output, and the precise moments in time when the returns will appear (De Giovanni, 2020). These three pillars are incorporated within the Waste Cascade Model, and their simultaneous analysis and evaluation lead to a proper assessment of CE systems.

Accordingly, CE systems must be carefully analyzed and inspected before diving into their implementation. The intention to adopt CE systems should be followed by a few key questions: How can firms benefit from CE principles? How should firms modify their business models to accommodate CE? How will the supply chain transform their contracts and relationships to also consider CE options? What are the main business opportunities behind CE? What are the main drivers and barriers to be considered when considering its implementation? Which real actions and policies can firms adopt to overcome the operational and logistical barriers linked to CE? These questions challenge firms and businesses, as well as consumers and governments. Hence, the entire ecosystem requires knowledge of CE's potential to appreciate the real business opportunities, the environmental implications, and the possible social benefits.

TARGET AUDIENCE

This book is a collection of case studies and business applications linked to CE. By reading this book, professionals, executives, and business leaders will obtain valuable information about how to implement CE systems, which barriers and challenges firms currently face, which opportunities CE actually offers, and possible links with smart cities. Furthermore, scholars and Ph.D. students will find models and systemic structures that follow the real business challenges linked to CE. The book also provides inspiration for future research projects on CE to better understand firms' strategies and policies. Finally, because it is composed of case studies, this book perfectly fits with graduate and undergraduate programs in the areas of operations management, supply chains, the circular economy, and sustainable strategies. Students will gain an understanding of how CE systems are really managed, as well as the managerial challenges firms face when integrating such systems.

ORGANIZATION

This book is organized into 11 chapters. Each chapter analyzes how CE systems are implemented in companies and districts by highlighting the business context, driving factors, such as consumers, governments, and stakeholders, opportunities and challenges, and potential links with smart cities. A brief description of each chapter follows.

Chapter 1 provides an analysis of how to concretely implement and manage innovative projects to shift from linear to circular supply chain management. The chapter analyzes the case of sustainable wooden packaging logistics implemented by Fercam Echo Labs, moving from recycling approaches to upcycling solutions for pallets and crates.

Chapter 2 is an application of Lithion's battery swapping service model and proposes it within the context of the circular economy. Lithion Power is India's leading provider of batteries as a service, supplying lithium-ion batteries for e-bikes and three-wheelers. Lithion collaborates with major operators and original equipment manufacturers (OEMs) to create an all-electric India in the next decade.

Chapter 3 proposes a link between the metaverse and the circular economy. These new technologies are very helpful in developing smart cities, with the goal of improving the quality of life considering sustainability. This link is introduced in detail and is examined within the framework of smart cities.

Chapter 4 analyzes the case of Fairphone, a Dutch company that makes ethical mobile phones relying on a circular business model. Given the complexity of the business model, Fairphone has created a unique network of stakeholders, with customers playing a pivotal role, each providing fundamental help. Given the innovative and holistic approach Fairphone has implemented, the company has had positive effects on the growing field of Smart Cities. The chapter is dedicated to the analysis of Fairphone's business case and the ongoing positive impacts the company could generate within Smart Cities.

Chapter 5 introduces a business model developed by Levi Strauss and Co. within the circular economy arena, with a focus on the materials report. The chapter shows how the company can become a benchmark for all firms in the textile industry by also exploiting digital technologies and leadership.

Chapter 6 introduces the case San Benedetto, which is a perfect example of a company that has developed a strong commitment to the environment since its origins. It has embraced the circular economy paradigm since it has closed the loop, reusing and converting returns.

Chapter 7 presents the case of CIER and how it approaches the circular economy. The chapter introduces values, drivers, stakeholders, and barriers linked to CE in the company and provides a set of indicators to measure the environmental and social impacts. The chapter ends with recommendations on the use of blockchain technology and smart city adoption.

Chapter 8 explores H&M and its impact on the textile value chain. The chapter demonstrates how developing eco-friendly materials, monitoring sustainable production, reducing carbon emissions in distribution, and promoting circular fashion can lead to success.

Chapter 9 elaborates on the application of CE systems in the textile industry, which is historically influenced by seasonality effects that induce a search for circular solutions independent of environmental constraints and regulations. This chapter shows that CE systems can create business opportunities as well as important cannibalization effects when consumers are confronted with new and used textile goods. Hence, it highlights how the use of digital technologies can help to support the mitigation of such issues.

Chapter 10 applies the theoretical concept of the waste cascade developed by De Giovanni and Folgiero (2022) to the Dell Reconnect case. It demonstrates how, starting from a theoretical model, the practices reveal the possible adoption of CE systems to gain only partial benefits from circularity. This implies that CE potential must be evaluated based on a singular case. Consequently, no perfect standard CE system exists. Rather, each case should be shaped by analyzing CE options, the returns' residual value, and involved stakeholders.

Chapter 11 explores the benefits that blockchain grants in the furniture sector to guarantee information transparency over the CE network, especially when it is integrated with Industry 4.0 technologies. The case demonstrates that using the blockchain is challenging due to the specialized facilities and the high level of machine automatization. However, it can represent a real opportunity for firms in the furniture sector to enable a smart and transparent CE system.

IMPACT ON THE FIELD AND CONTRIBUTIONS

This book proposes a set of case studies and business applications to shed light on the motivations, the drivers, the challenges, the value creation options, and stakeholder involvement when CE systems are in place. Furthermore, the various chapters link CE systems with possible applications to smart cities, taking them a step ahead in the management of businesses and CE processes in a revolutionized and digitalized future. All business cases proposed in this work are then developed in three directions. First, each chapter highlights the industrial and sectoral motivations for implementing CE systems. An analysis of the processes and the supply chain structure is then conducted to explore the entire ecosystem. For each of the business cases, the drivers, the motivations, the challenges, and the opportunities linked to CE systems are highlighted to shed light on the business motivations, as well as the environmental and social reasons, that support circular thinking. Finally, the cases are linked to the future development of smart cities as a launching point for imagining how the future of CE should be developed in a smart and digitalized framework, with a link to the Waste Cascade Model.

One interesting dimension that emerges from this book is that CE requires the full support of corporate strategy. In fact, CE can imply important modifications of firms' visions and targets, which imply a change in relationships with stakeholders, alignment with new regulations and legislation, as well as an analysis of new and emerging consumer needs. Future CE developments should be analyzed in these directions, considering that mega trends such as digital transformation, the COVID-19 pandemic, the Ukraine-Russia war, and the worldwide semi-conductor shortage can substantially complicate and/or accelerate the diffusion of CE systems. For example, the recovery materials from CE processes can compensate for part of firms' demand for virgin materials and, consequently, mitigate the shortage risks linked to raw materials.

Pietro De Giovanni
Luiss University, Italy

REFERENCES

De Giovanni, P. (2020). *Dynamic quality models and games in digital supply chains*. Springer Nature.

De Giovanni, P., & Folgiero, P. (2022). Strategies for the Circular Economy. Circular Districts and Networks. Giappichelli-Routledge. ISBN: 9788892123876.

De Giovanni, P., & Zaccour, G. (2022). A selective survey of game-theoretic models of closed-loop supply chains. *Annals of Operations Research*, *314*(1), 1–40. doi:10.100710479-021-04483-5

Ghisellini, P., Cialani, C., & Ulgiati, S. (2016). A review on circular economy: The expected transition to a balanced interplay of environmental and economic systems. *Journal of Cleaner Production*, *114*, 11–32. doi:10.1016/j.jclepro.2015.09.007

MacArthur, E., Zumwinkel, K., & Stuchtey, M. R. (2015). *Growth within: A circular economy vision for a competitive Europe*. Ellen MacArthur Foundation.

Acknowledgment

I would thank all the authors and the reviewers who have contributed to this book for their time and their dedication. A big thank goes to my wife, Maria, who supports me with patience and love. A special thank goes to my kids, Maria Greta and Riccardo, to whom I also dedicate this book, with the hope to enhance their circular thinking, inspire their future responsible behaviors, and stimulate their contributions for a greener and more sustainable world.

Chapter 1
Toward a Circular Supply Chain:
The Case of Fercam Echo Labs

Alessandra Cozzolino
https://orcid.org/0000-0003-4466-8769
Department of Management, Faculty of Economics, Sapienza University of Rome, Italy

EXECUTIVE SUMMARY

How circular economy systems truly work for firms around the world is at the beginning of knowledge development. As such, this chapter aims to provide an analysis of how to concretely implement and manage innovative projects to shift from a linear to circular supply chain management. The chapter analyses the case of sustainable wooden packaging logistics implemented by Fercam Echo Labs, moving from recycling approaches to upcycling solutions for pallets and crates. The development of circular supply chain management inside the circular economy can properly guide research and practitioners' efforts in the innovative logistics packaging management arena.

INTRODUCTION

The European Commission adopted the new "*Circular Economy Action Plan*"-CEAP (March 2020) as one of the main building blocks of the European Green Deal for the new European agenda for sustainable growth. The CEAP provides a future-oriented agenda for achieving a cleaner and more competitive

DOI: 10.4018/978-1-6684-5001-7.ch001

Europe in co-creation with economic actors, consumers, citizens and civil society organizations. In the document "*Circular economy action plan. For a cleaner and more competitive Europe*", packaging is placed among the "key product value chains", "where the potential for circularity is high", together with others: "electronics and Information and Communication Technologies"; "batteries and vehicles"; "plastics"; "textiles"; "construction and buildings"; "food, water and nutrients"[1]. The sustainability challenge especially posed by packaging value chains requires urgent, comprehensive and coordinated actions, which form an integral part of the European sustainable product policy framework and industrial strategy and contribute to the response to the climate emergency[2].

The presence of packaging along the supply chains in fact is pervasive, both as a product itself and as a combination of product-packaging (Silva and Pålsson, 2022; Massaroni and Cozzolino, 2021; Cozzolino, 2021a). Due to its pervasiveness along supply chains, packaging carries great importance in achieving sustainable goals and targets (Fitzpatrick et al., 2012), supported through the development of guidelines, standards, and scorecards that should be applied over the entire packaging life cycle – from production, through to distribution and transport, to use and disposal (Kozik, 2020). It is precisely the packaging "logistics" processes that make product packaging possible, flowing through the entire supply chain and defining interaction with the physical environment and the socio-economic context (Cozzolino, 2021a; Vernuccio *et al.*, 2010). Thus, the idea of "sustainable" packaging logistics has been growing in academia and professional contexts (Cozzolino, 2022; Massaroni and Cozzolino, 2021; Cozzolino, 2021a). The concept of sustainable packaging logistics is connected with a strategic, systemic and holistic view, going beyond a formal system (i.e., accounting, social and environmental responsibility), imposed by rules and regulations, to a sustainability that works "toward a triple helix for value creation, a genetic code for tomorrow's capitalism, spurring the regeneration of our economies, societies, and biosphere" (Elkington, 2018), inside the circular economy perspective.

The circular economy perspective, promoted by the Ellen MacArthur Foundation[3] is a "systems solution framework" that tackles global challenges, such as climate change, biodiversity loss, waste, and pollution, aimed at retaining as much value as possible of products, parts and materials. It is increasingly recognized as a better alternative to the dominant linear economic model (focused on take, make, and dispose) promoting products and services traded in loops or cycles (Ghisellini et al., 2016; Elia et al., 2017; Geissdoerfer et al., 2017; Kirchherr et al., 2017). Companies are showing an increasing interest for this new economic model (Elia et al., 2017), improving new business models to better ensure a constant flow of services and goods without the need for new materials or raw materials through different "value circles"[4]. Specifically, the integration of the approach of the circular economy into supply chain management offers a new and compelling perspective to the supply chain sustainability domain (Genovese et al., 2017; Nasir et al., 2017). Consequently, there is increasing research interest in this viewpoint by many authors (Ying and Li-jun, 2012; Aminoff and Kettunen, 2016; Batista et al., 2018; Bressanelli et al., 2019; De Angelis et al., 2018; Govindan and Hasanagic, 2018; Howard et al., 2019; Liu et al., 2018), as underlined by Farooque et al. (2019) in their literature review paper on circular supply chain management, which proposes great potential in this research direction.

However, research is still at a nascent stage and great potential would be realized in this direction (Farooque et al., 2019). Inside this new framework concreate solutions may be developed but are still at a beginning phase stage. The development of circular supply chain management inside the circular economy can properly guide research and practitioners' efforts in the packaging management arena, both in designing and adopting ad hoc practices along the supply chain (Meherishi et al., 2019). In fact, packaging plays a fundamental role in circular economy activities, including packaging return, reuse, recycle, upcycle and disposal (Cozzolino, 2022; Silva and Pålsson 2022; Sung et al., 2021; Meherishi et al., 2019). This gap is underlined also by the European Commission[5] in order to ensure that all packaging on the EU market is reusable or recyclable in an economically viable way by 2030, at least with a focus on reducing (over)packaging and packaging waste and driving design for re-use and recyclability of packaging.

Along this research gap, considered a current and burning topic both by academicians and practitioners, the paper aims at investigating the following research question:

RQ – What initiatives are implemented by companies to realize packaging logistics innovations to shift from a linear to a circular supply chain perspective?

To answer the research question, the rest of the chapter is organized as follows: Section 2 presents the theoretical background considering the intersection among the areas of packaging, logistics and sustainability; Section 3 describes a case study on sustainable packaging logistics from a circular supply chain management perspective, presenting the concrete project developed by Fercam Echo Labs; in Section 4, some insights are proposed, and conclusion is reported in Section 5.

PACKAGING, LOGISTICS AND SUSTAINABILITY: FROM A LINEAR TO A CIRCULAR SUPPLY CHAIN PERSPECTIVE

The integration of the circular economy into supply chain management has been termed "circular supply chain" in the literature (Farooque et al., 2019; Batista et al., 2018; Nasir et al., 2017; Genovese et al., 2017), and, in particular, Farooque et al. (2019 p. 884) defines "circular supply chain management" as "*the integration of circular thinking into the management of the supply chain and its surrounding industrial and natural ecosystems. It systematically restores technical materials and regenerates biological materials toward a zero-waste vision through system-wide innovation in business models and supply chain functions from product/service design to end-of-life and waste management, involving all stakeholders in a product/service lifecycle, including parts/product manufacturers, service providers, consumers, and users*". The purpose of circular supply chain management is to lead away from a linear supply chain towards a circular supply chain. A linear supply chain extracts resources from the geosphere and the biosphere and disposes of products, packaging materials, and wastes from multiple supply chain stages, while a circular supply chain improves environmental performance by recovering as much as possible value from waste in strong collaboration with other organizations (Farooque et al., 2019), within the industrial or other sectors (Weetman, 2017).

The circular supply chain concept relies on the concept of Closed-Loop Supply Chains (CLSC), that are characterized by the integration of forward and reverse flows in a unique system that consists of recycling, reusing, remanufacturing, and refurbishing, along with the traditional sales and service forward activities (Jalali et al., 2022; De Giovanni and Zaccour 2019; De Giovanni and Zaccour, 2014; Quariguasi Frota Neto et al. 2010; Guide and Van Wassenhove, 2006). Previous research has explored CLSCs covering different aspects and perspectives of this field (Souza, 2013; Govindan et al., 2015; De Giovanni, 2016; Guo et al., 2017; Ramani and De Giovanni, 2017; Genc and De Giovanni, 2020). Farooque et al. (2019) underline as the extent of value recovery in a closed loop supply chain is often limited because the efforts are restricted within the original supply chain (producer's supply chain) and do not include secondary supply chains and/or involve new channel members (Moula et al., 2017). Along this consideration, an interesting area of research seems to emerge considering the packaging, placed among the "key product value chains" by the European Commission in the "Circular Economy Action Plan" (Cozzolino, 2022).

Packaging fulfils fundamental role along the supply chain in assuring the availability of the right product, in the right quantity, in the right condition, in the right place, at the right time, to the right customer, at the right price (Shapiro and Heskett, 1985). Such "rules of availability" lead, in today's competitive landscape, to the formulation of original packaging for "complete logistical solutions" (Chapman et al., 2002), oriented towards solving "complex processes" involving different actors, many functions, and different requirements and conditions to pay attention to (Hellström and Saghir, 2007). In this sense, packaging is a complex system, so Twede (1992) refers to the "packaging system" as composed of three levels of packaging intimately related to logistics, in primary packaging (or "sales packaging", or "consumer packaging"), secondary packaging (or "group packaging", or "distribution packaging"), and tertiary packaging (or "transport packaging"). This system explicitly recognizes packaging as a hierarchical system, the performance of which is affected also by the interactions between levels, and not only by the performance of each single packaging level (Hellström and Saghir, 2007).

Many scholars have highlighted the principal logistical functions required of this packaging system along the supply chain, in pursuit of optimum efficiency, cost reduction, time savings, and qualitative service performance (Friedman and Kipnees, 1977; Harkham, 1989; Ebeling, 1990; Paine, 1991; Twede, 1992; Twede and Parsons, 1997; Lambert et al. 1998; Saghir and Jönson, 2001; Soroka, 2002; Hellström and Saghir, 2007; Robertson, 2006; Williams et al., 2008; Yam, 2009; Vernuccio et al., 2010). In synthesis, they are mainly the following (Vernuccio et al., 2010): protection and conservation (a); transport, handling and storage (b); and information (c). Protection and conservation (a) refers to the primary key function of primary packaging. Packaging, on the one hand, safeguards the physical, thermal and chemical-bacteriological integrity of the product; on the other hand, it prevents injury to the user or damage to the external environment in which the product is used, in the case of hazardous or potentially polluted content. Transport, handling and storage (b) refers to all the levels of packaging (primary, secondary, and tertiary) that collectively permit and facilitate various operations throughout all the phases of production, distributive and consumption processes along the supply chain. Information (c) refers to the communicative function of packaging. There could be messages incorporated into the materials used that deliver practical instructions to those involved in the movement of goods through the distribution chain and information about the contents.

Referring to the logistics functions, the logistics processes that sustain the cycle, for the forward and return, should be considered (Cozzolino, 2021a). Even more alongside the direct/forward logistical flow (from upstream to downstream) is the return flow of packaging (and packaged product) – reverse/return logistics – which has attracted more attention in recent years, mainly for sustainable reasons (Meherishi et al., 2019). Packaging is an integral part of the logistical system (Kirwan and Coles, 2011). Many authors have emphasized the close relationship between the concepts of "logistics" and "packaging" (García-Arca et al., 2014; Azzi *et al.*, 2012; García-Arca and Prado-Prado, 2008; Hellström and Saghir, 2007; Verghese and Lewis, 2007; Saghir, 2002; Lockamy, 1995; Twede, 1992), which focuses on the "synergies" achieved by integrating logistics and packaging with the potential of increased supply chain efficiency and effectiveness (Vernuccio *et al.*, 2010). The link between logistics and packaging was initially summarized in the expression "logistical packaging", which was used by academics referring to the personalization of packaging useful for logistics (Paine 1991; Twede 1992; Twede and Parsons, 1997). This concept does not communicate the full potential of the link between logistics and packaging: it was considered more appropriate to use the expression "packaging logistics", which, in addition to focusing on the interface between packaging and logistics systems, recognizes the interdisciplinary nature of the packaging itself and fully enhances the interaction among packaging, logistics and marketing decisions (Saghir, 2004). Thus, the expression "packaging logistics" refers to the integration of packaging design with logistics management, with a particular emphasis on strategic aspects (García-Arca et al., 2014; Saghir, 2002; Hellström and Saghir, 2007).

Packaging (logistic packaging and packaging logistics) has received a huge consideration in the literature on logistics and supply chain management due to its impact on sustainability. Packaging may carry great importance in achieving sustainable goals and targets (Fitzpatrick et al., 2012), since the first definitions express the concept of "sustainable packaging". Sonneveld et al. (2005) present the work of James et al. (2005), inside the Sustainable Packaging Alliance, defining "sustainable" packaging that is safe (nontoxic to humans and ecosystems), cyclic (recyclable or compostable), efficient (minimum use of materials and energy) and effective (adds economic and social value). Based on this definition, other authors and institutions refined it with an emphasis on the "life cycle" (Meherishi et al., 2019; Fitzpatrick et al., 2012; the Sustainable Packaging Coalition[6]; the European Organization for Packaging and the Environment[7]).

The definitions of sustainable packaging are in line with the perspective of the circular economy and circular supply chain, also if it is still not easy to univocally define it in concrete solutions due to the multitude of criteria that should be considered, a large variety of packaging materials, the dynamic development of the industry, and different actors involved throughout the whole life cycle of the product and of the packaging-product combination (Kozik, 2020; Lindh et al., 2016). In a general view, sustainable packaging logistics, compared to conventional one, meet higher environmental, economic and social standards, have better performance and quality features, and, at the same time, bring new possibilities in the field of recovery and waste management (Kozik, 2020). In this way, packaging "upcycling" presents a promising alternative to mass production and consumption based on the use of virgin materials, slowing the material cycles through a circular perspective (Sung et al., 2021). Upcycling means in fact the reuse of discarded objects or materials in such a way as to create a product of higher quality or value than the original (Wegener, 2016). The term upcycling originated in the 1990s (Bridgens et al., 2018), but research on upcycling, however, is still in the developing stages (Sung et al., 2021).

Upcycling has the potential to transform the way we consider individual products as assemblages of functional component modules with multiple lifespans, rather than complete stand-alone objects with singular finite lives (Richardson, 2011). Despite variations among definitions, there are two dominant viewpoints in the analyzed publications, according to Sung (2015), one based on material recovery, of which the major aim is to maintain value and quality of materials safely in their second life and beyond by the improved recycling or remanufacturing; the other focused on product (re)creation for higher values and qualities by transforming, repurposing or refashioning waste or used materials/products either by companies or by individuals. For industrial upcycling, both as upgraded recycling and as remanufacturing, specialist skills, equipment, tools, space and time are needed. The area of upcycling as a concrete application is relatively new and unexplored, and industrial practices – who is doing what, when, where and how, and how (un)/successful it is – remain largely unknown (Bridgens et al., 2018).

Especially for some specific packaging, based on their material, the upcycling solution might be interesting to be analyzed in practice. For example, for wood packaging, such as pallets and crates, the recycling process typically involves breaking down the original material and making it into something else, for example, chipboard or medium-density fiberboard (MDF) panels (Ali et al., 2013). In this specific case of wooden pallets and crates, the shredding process compromises the essence, and producing new boards necessitates deforestation. And if we especially consider wooden pallets, that they are both expensive to buy and expensive to throw away when they brake (Adsavakulchai, 2021), upcycling preserves the integrity of the material (and is totally energy saving), creating new products, often with a greater value than the original object, and the process of creative reuse interrupts the deforestation process (Ali et al., 2013).

Moreover, considering pallets management in a circular perspective permits to focus on an extremely crucial type of packaging for transport and logistics along the entire supply chain since they are key elements of almost every global supply chain carrying approximately 80% of all world trade, with the potential to make the difference in terms of sustainability (Tornese et al., 2021).

Alongside the mentioned theoretical considerations, the empirical part of this paper focuses on a case study. Since the nature of the research question requires an exploratory approach, a case study method seems to be the most suitable for this research (Yin, 2003). The case study methodology is well recognized as a valid approach through which to deepen understanding of a phenomenon that is still in development and/or for which the dimensions have not yet fully explained (Yin, 2003; Eisenhardt, 1989). In particular, logistics and supply chain researchers have also advocated the use of case study research as an approach to scientific inquiry to provide detailed explanations of best practices (Ellram, 1996). In accordance, the specific case has been chosen because it is of interest (Stake, 2005) and its relevance for theoretical reasons (Eisenhardt and Graebner, 2007).

The case study included in this chapter focuses on upcycling solutions for wooden transport (or tertiary) packaging, including pallets and crates, in a circular supply chain perspective. Wooden pallets are load support that are indispensable for the logistics, transport, handling, and storage of goods in almost every industrial sector, while wooden crates are logistical boxes useful for moving highly value products, in this paper, in particular, they are used for fine-art transport. Upcycling for pallets and crates offers a whole new life for discarded materials, taking a position in alternative to unusability or recycling. Both of them are managed by Fercam, primary Italian family business in transport and logistics, that has become one of the main logistics service providers- LSPs in Europe, inside its business activities. They are readjusted as long as it is possible, and in alternative they are completely recycled. But in October 2021 inside the distribution center of Rome a pilot-project has been started along the idea of the upcycling solution for a better circularity. The choice of this case derives from the desire to follow the development of that pilot-project, from the starting idea to the complete implementation of the solution, and even beyond what was initially imagined: as they declare "*As from a single small drop bigger and bigger waves are generated, from this first limited initiative we have begun to broaden our range of action. We understood that we could have a significant and positive impact, leaving a footprint we can be proud of, in terms of environmental and social sustainability. Thus, was born Fercam Echo Labs*"[8]. The case in this sense it's a one-of-a-kind case at the moment and it may represent an innovative best practice in the sector.

THE CASE OF FERCAM ECHO LABS[9]

Fercam Echo Labs is a nonprofit social enterprise: a "*permanent laboratory*" that was born in December 2021 and works on a more sustainable future for people and the environment. Its mission is to create networks and synergies between the Corporate Social Responsibility programs of its clients and partners, which thanks to the active participation in the laboratory are able to carry out projects that individually they could not have carried out, preferably starting from the territory in which they manage their activity. It proposes itself to them as General Contractor for the realization of sustainable projects. Fercam Echo Park, just inaugurated in May 2022, is one of those projects and represents the starting idea since its beginning in October 2021.

The Echo Park is an outdoor structure inside the area of the Fercam distribution center of Rome, where they have created a break area for lunch or coffee available to employees and indirect collaborators. In this area, wooden pallets were used for furnishings and accessory structures, valorizing the wood material into new life through upcycling solutions. The nonprofit association Linaria was identified as a project partner to design and guide the creation of furnishings and accessory structures starting from wooden pallets, together with flowering plants and plug hotels to welcome solitary bees, to promote biodiversity and disseminate knowledge of urban nature, with a concrete and widespread environmental action of the regenerative nature. The woodworking activities were a learning opportunity for a group of political refugees, and with the hours of training in eco-carpentry led by the Laboratorio Linfa experts, the participants in the project were awarded a certificate, which is useful for entering the European world of work, inside the framework of the inclusion and integration system. In fact, this initiative also sees the involvement of the social cooperative Medhospes, the managing body of the SAI project (hospitality and integration system) for the Roma Capitale local authority. Two obsolete containers, otherwise destined for disposal for the recovery of iron, have been modified, repainting them and obtaining openings to make large windows to create a covered structure to complete the break area. Inside the structure, other complementary sustainable activities have been realized, for example, with vending machines for the consumption of healthy, organic, zero-kilometer or artisanal products, plastic compacting machines for recycling PET bottles, coffee cups, and the recycling of coffee grounds.

Moreover, in this dedicated break area and inside the Company offices, bookcases will be installed to encourage reading with voluntary and free bookcrossing initiatives, leaving books available to anyone who wants to read them, especially long-haul line drivers, during their long stops, between one stage and another. The book will then be able to travel with them, in company branches or wherever they want to leave it. The bookcases will be made by upcycling the wooden crates built for the transport of works of art, utilized by Fercam Fine Art. In Fercam Fine Art, the internal carpentry activity is one of the pillars of artwork handling operations, independently producing all of the crates for transport activities. The legislation that regulates the transfer of cultural assets requires the precise rules for packaging materials be respected, so the company uses valuable multilayer wood panels to manufacture the crates. Whenever possible, they readjust them for subsequent transport, but the production needs of new crates still outweigh the opportunities for reuse. It is a noble wood, both for the nature of the material and for its intended use; therefore, it has always been the company priority to find the right methods to reutilize it, possibly for equally noble purposes. Additionally, they have already reused the wood from these crates to make kennels for animals, the last of which was recently donated to municipal kennel.

Fercam Echo Labs is realizing, with wooden pallets and crates upcycling, among others, projects of sustainability in each of its three dimensions—economic, environmental, and social—simultaneously and alongside the supply chain, from a circular perspective. Upcycling solutions may not cover all volumes of wood packaging, so this solution could be proposed, in combination with recycling initiatives, that Fercam has already activated. For example, at Fercam in Rome, an agreement was already in place with

Mediaservice Recycling for the recycling of all materials used in the transport and distribution of goods, from cartons to pallets and packaging. A new agreement with Saviola has been created especially for crates for works of art transport that are no longer reusable. There is no unique best way to go forward within a circular supply chain, but rather through a combination of concrete activities that have the same aim of retaining as much value as possible of products, parts and materials in loops or cycles.

The upcycling idea of the Fercam Echo Labs was developed starting from a simple need of the Fercam Rome distribution center: creating a relaxed area available to employees and collaborators in an eco-sustainable manner. Fercam Echo Park was the pilot project, and the idea was born from Fercam S.p.A. Regional Manager, Dino Menichetti, who is now also Fercam Echo Labs President. He declared: "*In [Fercam] corporate activities, Corporate Social Responsibility is of great importance. With Echo Labs, we wanted to go beyond the Fercam business to establish a network of collaborations and nonprofit partnerships. The synergies created between the CSR programs of the various entities involved will allow each to have an even more significant positive impact, under the aegis of the 17 Sustainable Development Goals of the UN 2030 Agenda*"[10]. Many other innovative projects are in the developing phase with Fercam, and also with other companies and institutions in the territory, for example schools.

This specific case is not the only one Fercam carries out in these recent years especially in the direction of a more complete sustainability.

INSIGHTS FROM THE CASE

Theoretical Implications

Some theoretical considerations have emerged from the description of the case.

First, the case of Fercam Echo Labs proposes concrete and innovative solutions for an extremely crucial type of packaging for transport and logistics along the entire supply chain, namely pallets.

Pallets are key elements of the global supply chains and are one of the most used returnable transport items, carrying approximately 80% of all world trade (Tornese et al., 2021). The pallet supply chain is a complex product network, and the way pallets are managed throughout their lifecycle phases produces a notable difference in terms of environmental and economic impacts

(Tornese et al., 2021; Gnoni et al., 2018). Analysts estimate that there are approximately five billion pallets in use worldwide, of which two billion are in the United States alone; approximately 90% of the pallets are made of wood, and the rest are made of plastic, metal, or cardboard[11]. Analysts also speculate that the market will remain a hot one: in 2020, the global pallet market size was valued at 78 billion U.S. dollars and is expected to reach over 110 billion U.S. dollars by 2027[12]. Complexity also arises according to the use of pallets, as they could be "disposable" pallets, if used only once, also known as nonreusable or disposable, "reusable" pallets, intended to be used several times, also called multirotation, "internal" pallets whose use is limited to a single company or a closed distribution system, pallet "for exchange use", which, on the basis of a mutual agreement, can be exchanged for an identical pallet in a "sharing" perspective[13]. Inside the framework, adapted by Kirchherr et al. (2017) from Potting et al. (2017), the pallets may represent very interesting objects for increasing circularity in the 3 different strategic areas that lead from a linear economy to a circular economy: the useful application of materials, the extended lifespan of products and parts, and smart product use and manufacture.

Second, the case of Fercam Echo Labs refers to upcycling solutions for wood, not as the opposite of recycling, but "in combination with".

Wood is very interesting and promising from a circular supply chain perspective (Ali et al., 2013). It can be reused, readjusted, recycled and upcycled, in a full portfolio of opportunities inside the circular economy. In particular, between recycling and upcycling the difference is on the prevention of the integrity of the material and this point has a consequence also on the supply of new materials and on the deforestation activities (Ali et al., 2013). According to Rilegno[14], the Italian national consortium that deals with the collection, recovery and recycling of wood packaging inside the Conai system (Italian National Packaging Consortium), wooden packaging is among the most used for the transport of goods including especially pallets and crates (made with continuous covering in boards or panels), but also cages (made with a discontinuous board coating), coils (for winding electric, metal or fiber optic cables and ropes), food boxes and corks. Rilegno in Italy has the task of guaranteeing the achievement of the objectives set by law for the overall recovery of postconsumer wood packaging from, both separate collection carried out by the municipal administrations, and from collection carried out by private operators from industries and large-scale distribution. Inside the circular supply chain of wood packaging, in 2019, 2,999,772 tons of packaging were used in the Italian national territory, of which 1,425,104 were new pallets, 847.091 were returned pallets, and 428.250 were industrial crates and cages. The share of wood packaging recovered from the released for consumption is 64.68%; 1,841,065 tons of wood were collected and recycled by Rilegno.

Third, the case of Fercam Echo Labs represents the birth of an organizational solution: a nonprofit social enterprise, to better propose – inside the economic context of the logistics services – concrete projects toward a circular supply chain, also with strategic alliances.

The idea of the structural form of the social enterprise has attracted the attention of policy-makers, practitioners, and academia around the world. Social enterprises are emblematic of economy-wide increases in activity at the interface between business and charity, as corporations increasingly engage in social responsibility-related activities, and nonprofits increasingly engage in commercial activities to complement their primary, philanthropic sources of funding (Battilana and Lee, 2014). Social enterprises may become a "platform" for innovating in the direction of sustainable business models. The concept of a platform represents a space/interface in which different actors, skills, abilities, resources, knowledge, objectives, or needs could be easily finalized in a concrete interaction, which otherwise would be much more difficult to achieve. A platform may realize this by facilitating the matching of interests, by creating communication opportunities among different actors, by setting up standardized technological interfaces, and by building cross-functional teams (Ansell and Gash, 2018). According to this, main platform characteristics emerge into the platform "framework", including activities, resources and people, which may be better organized for a common objective. This framework is quite stable over time in terms of organizational structure, however, the activities, resources, and people who it comprises may be easily "reconfigurable" (Ansell and Gash, 2018). The platform is an active support structure that facilitates "reconfigurability" (Ciborra, 2009). Reconfigurability facilitates the emergence of an "ecosystem" of complementary activities, resources, and people (Jacobides et al., 2018; Gawer and Cusumano, 2014). Going deep into the analysis by focusing on ecosystems (and on networks and systems), especially in a service context, would enable practitioners to extend their perspective to deal with increasingly complex business settings and cocreate value (Barile et al., 2016).

Fourth, the innovative cases and the foundation of social enterprise are taken from the experiences of Fercam, as a logistics services provider company.

Logistics services are crucial in the implementation of sustainability in each of its three dimensions—economic, environmental, and social. This is, simultaneously, alongside the supply chain, and because of the high level of outsourcing of logistics activities at the global level, logistics service providers (LSPs) play a "pivotal" role in sustainable supply chain management, both as companies and as service providers for other organizations. LSPs can then assume a critical role in supply chain orchestration and management toward sustainability, having the capabilities to develop solutions for more sustainable supply chains (Colicchia et al., 2013; Centobelli et al., 2017; Anttonen et al., 2013). They can develop solutions by investing in sustainability through innovation, which they can incorporate into their own organization's products and processes, as well as in their relationships with suppliers and customers. Moreover, as they serve a large number of customers and partners in diverse sectors and geographical areas, both in the manufacturing and commercial fields, the positive effects of their activities may easily multiply (Cozzolino, 2009; Massaroni and Cozzolino, 2012; Massaroni et al., 2016). They will be in a good position to "orchestrate" the complex logistics networks where they operate toward a more sustainable supply chain (Christopher, 2005). The term metaphorically recalls the role of the symphonic orchestra "maestro" who composes the symphony, defining the timing and synchronisms of each musical element (as composer), and directs the individual contributions in the execution of the planned harmony (as conductor). The metaphor of the orchestra maestro can be transferred from the individual subject to organizations, referring to a general combination of complementary resources for a common project (Cozzolino and Vakharia, 2007; Cozzolino 2009; Cozzolino, 2021b). It describes well the capabilities and the strategic position LSPs may play for their supply chains: the contribution of each element is an indispensable component of overall performance, and they are a strategic multi-contact points for clients, suppliers and partners in the direction of sustainable processes and initiatives within a circular perspective.

These considerations emerging from the case may inspire future research and case investigations in the same domain.

Managerial Implications

Numerous important issues are combined in defining the complexity of the topic investigated in this chapter and its relevance in the real context. However, the analysis of the case would suggest that some may be key challenges for managerial considerations. They could be useful to inspire certain instructions/strategies on how similar initiatives can be empirically

Table 1. Key-challenges for strategies and managerial considerations

Key-challenge	Current setting (case)	To-be-setting (general)	How to get there?
Fostering supply chain collaboration	"Permanent laboratory" mainly involving company workers, no-profit associations and professionals, refugees ...	Inter-organizational and cross-sector "learning laboratory" potentially involving all supply chain actors and stakeholders	Exploiting complementariness (of competencies, skills, know-how, knowledge, resources ...)
Valorizing a "portfolio" of options inside the sustainable strategy	To reuse, repair, recycle and upcycle are all concrete options utilized inside the business strategy	"Portfolio" of complementary options for the company sustainable strategy along the supply chain	Developing multi-criteria methods and operationalizing measures to sustain decision-making process
Managing simultaneously economic, environmental, and social sustainability	Integration of the environmental sustainability with economic and social dimensions	Composition of complete service-product-process solutions to simultaneously satisfy sustainability dimensions	Innovating in the areas of integration/overlapping among dimensions of sustainability

improved in a supply chain perspective. To be sustainable in only one point of the supply chain it is not sustainable (Massaroni et al., 2015): as Preuss (2005) stated, lasting sustainability can be ensured only if companies move beyond their own confines and establish relationships to fully benefit from sustainability management. A supply chain perspective then is no longer just an exciting opportunity, it has become a "survival skill" (Rezapour et al., 2014), especially in a context where competition no longer takes place among single companies, but among supply chains (Christopher, 2005).

The main key-challenges are the following, as also summarized in Table 1:

(a) Fostering supply chain collaboration;
(b) Valorizing a "portfolio" of options inside the sustainable strategy;
(c) Managing simultaneously the three dimensions of sustainability—economic, environmental, and social.

The first (a) guarantees that the right supply chain perspective of circularity is concretely improved not only in one point of the supply chain and by one single actor belonging to it, but considering potentially all possible actors upstream and downstream from the focal company and also its horizontal partners, its stakeholders, its competitors, and even "simply" citizens, communities and territories. The pilot-project starts as mainly an internal work, but it expands quickly in a "permanent laboratory" that become a "learning laboratory" involving in the collaboration not only company workers, but also different no-profit associations and professionals that are expert in environmental and social initiatives, and political refugees (as workers to be formally formed). The pilot idea, starting as a company internal project, has developed in a wider project through the foundation of the nonprofit social enterprise. Inside this new form, the same project will be replicated also for other areas inside and outside the company, collaborating with schools, public areas and other subjects on the territory; and other projects may be coherently developed. Alongside this direction, it is needed to exploit the complementarities among different actors.

The second (b) inspires to consider multiple initiatives for the composition of a complete "portfolio" of options inside each company sustainable strategies. In particular, according to the three principles of the circular economy proposed by the Ellen MacArthur Foundation[15], portfolio initiatives should eliminate waste, circulate products and materials (at their highest value), and regenerate nature. Alongside reuse, repair, recycle and upcycle are all concrete options utilized inside the business strategy for wooden pallets and crates management for the company. Wooden pallets could be circulated by being maintained, shared, reused, repaired, and, as a last resort, recycled. While recycling is the option of last resort because it means the embedded value in products and components are lost, it is vitally important as the final step that allows materials to stay in the economy and not end up as waste. These initiatives are complemented by many other innovative sustainable initiatives inside the sustainable strategy of the company, for example in transport and energy saving. Alongside this direction, it is needed to develop multi-criteria methods and operationalizing measures (for inputs, processes, outputs, impact, …) to sustain decision-making process for the portfolio composition.

The third (c) ensures guiding the innovations into a more complete approach of a fully sustainability (economic, environmental and social) in the next steps. The analysis of complex interactions among economic, environmental and social aspects has provided great opportunities for companies and supply chains (Massaroni et al., 2015), underlying the need for systems approaches to the complexity of sustainability issues (Barile and Saviano, 2018). The pilot-project included since its beginning the three dimensions of sustainability, and along the working-in-progress, the area of intersection among dimensions has increased, becoming synergic (economic > value of the pallets and their upcycling products, at least; environmental > wood material savings and bees hotels, at least; social > formative activities involving refugees and benefits for workers using the break area and the library for sharing books, at least). This could be empowered applying a systemic approach such as the results is more than the sum of the single parts. Alongside this direction, it is needed to innovate in the design of areas of integration/overlapping among the dimensions of sustainability.

These implications emerging from the case may offer a concrete roadmap for managers that wish to follow same paths.

CONCLUSION

Within the research stream of circular supply chain management, the debate on the impact of logistical packaging, and especially packaging logistics on sustainability, has shifted toward a holistic discussion of the impact of the packaging life cycle throughout the supply chain. Exploring how circular economy systems truly work for firms around the world is at the very beginning of knowledge development. As such, the chapter aims to provide an analysis of how to implement and manage innovative projects to shift from a linear to a circular economy, moving from recycling approaches to upcycling solutions in a circular supply chain perspective. In particular, the case presents the empirical experience implemented by Fercam Echo Labs from the starting pilot project to the foundation of the nonprofit social enterprise: ideas are described, including how they have been developed, and what initiatives have followed, including structural organizational decisions and strategic alliances. The case study represents an innovative best practice inside the sector and provides preliminary concrete evidence that could inspire research and practice. This study is a starting point for future steps on this topic, at both the theoretical and empirical levels.

ACKNOWLEDGEMENTS

The Author would like to thank Dino Menichetti for the opportunity to know about the Fercam Echo Labs project. The Author takes exclusive responsibility for opinions that are expressed in this chapter, and that do not necessarily reflect the view of the described organization.

REFERENCES

Adsavakulchai, S. (2021). Optimization of pallet management for transportation efficiently. *Journal of Research in Administrative Sciences, 10*(1), 19-22.

Ali, N. S., Khairuddin, N. F., & Zainal Abidin, S. (2013). Upcycling: Re-use and recreate functional interior space using waste materials. In *DS 76: Proceedings of E&PDE 2013, the 15th International Conference on Engineering and Product Design Education* (pp. 798-803). Academic Press.

Aminoff, A., & Kettunen, O. (2016). Sustainable supply chain management in a circular economy—towards supply circles. In *International Conference on Sustainable Design and Manufacturing* (pp. 61-72). Springer. 10.1007/978-3-319-32098-4_6

Ansell, C., & Gash, A. (2018). Collaborative platforms as a governance strategy. *Journal of Public Administration: Research and Theory*, *28*(1), 16–32. doi:10.1093/jopart/mux030

Anttonen, M., Halme, M., Houtbeckers, E., & Nurkka, J. (2013). The other side of sustainable innovation: Is there a demand for innovative services? *Journal of Cleaner Production*, *45*, 89–103. doi:10.1016/j.jclepro.2011.12.019

Azzi, A., Battini, D., Persona, A., & Sgarbossa, F. (2012). Packaging design: General framework and research agenda. *Packaging Technology & Science*, *25*(8), 435–456. doi:10.1002/pts.993

Barile, S., Lusch, R., Reynoso, J., Saviano, M., & Spohrer, J. (2016). Systems, networks, and ecosystems in service research. *Journal of Service Management*, *27*(4), 652–674. doi:10.1108/JOSM-09-2015-0268

Barile, S., & Saviano, M. (2018). Complexity and sustainability in management: insights from a systems perspective. In *Social dynamics in a systems perspective* (pp. 39–63). Springer. doi:10.1007/978-3-319-61967-5_3

Batista, L., Bourlakis, M., Smart, P., & Maull, R. (2018). In search of a circular supply chain archetype–a content-analysis-based literature review. *Production Planning and Control*, *29*(6), 438–451. doi:10.1080/09537287.2017.1343502

Battilana, J., & Lee, M. (2014). Advancing research on hybrid organizing–Insights from the study of social enterprises. *The Academy of Management Annals*, *8*(1), 397–441. doi:10.5465/19416520.2014.893615

Berg, P., Feber, D., Granskog, A., Nordigarden, D., & Ponkshe, S. (2020). *The drive toward sustainability in packaging—beyond the quick wins.* McKinsey & Company. https://www.mckinsey.com/industries/paper-forest-products-and-packaging/our-insights/the-drive-toward-sustainability-in-packaging-beyond-the-quick-wins

Bressanelli, G., Perona, M., & Saccani, N. (2019). Challenges in supply chain redesign for the Circular Economy: A literature review and a multiple case study. *International Journal of Production Research*, *57*(23), 7395–7422. doi:10.1080/00207543.2018.1542176

Bridgens, B., Powell, M., Farmer, G., Walsh, C., Reed, E., Royapoor, M., Gosling, P., Hall, J., & Heidrich, O. (2018). Creative upcycling: Reconnecting people, materials and place through making. *Journal of Cleaner Production*, *189*, 145–154. doi:10.1016/j.jclepro.2018.03.317

Centobelli, P., Cerchione, R., & Esposito, E. (2017). Environmental sustainability in the service industry of transportation and logistics service providers: Systematic literature review and research directions. *Transportation Research Part D, Transport and Environment*, *53*, 454–470. doi:10.1016/j.trd.2017.04.032

Chapman, R. L., Soosay, C., & Kandampully, J. (2002). Innovation in logistic services and the new business model: A conceptual framework. *Managing Service Quality*, *12*(6), 358–371. doi:10.1108/09604520210451849

Christopher, M. (2005). *Logistics & supply chain management.* Pearson Uk.

Ciborra, C. U. (2009). The platform organization: Recombining strategies, structures, and surprises. In Bricolage, Care and Information (pp. 134-158). Palgrave Macmillan.

Colicchia, C., Marchet, G., Melacini, M., & Perotti, S. (2013). Building environmental sustainability: Empirical evidence from Logistics Service Providers. *Journal of Cleaner Production*, *59*, 197–209. doi:10.1016/j.jclepro.2013.06.057

Cozzolino, A. (2009). *Operatori logistici* (Vol. 87). Wolters Kluwer Italia.

Cozzolino, A. (2021a). *The logistics functions of packaging: sustainable innovations toward a sustainable supply chain.* Sinergie-SIMA 2021 Conference "Leveraging intersections in management theory and practice", Palermo, Italy.

Cozzolino, A. (2021b). Platforms enhancing the engagement of the private sector in humanitarian relief operations. *Sustainability*, *13*(6), 3024. doi:10.3390u13063024

Cozzolino, A. (2022). From linear to circular supply chain: insights from case study in sustainable packaging logistics. In *Proceedings of the 7th International Conference on New Business Models: Sustainable Business Model Challenges: Economic Recovery and Digital Transformation*. LUMSA University.

Cozzolino, A., & Vakharia, A. J. (2007). 4PL-Providers as "Orchestrators" of Logistics Networks. *Proceedings of the Special Conference of Strategic Management Society (SMS), New Frontiers in Entrepreneurship: Strategy, Governance, and Evolution.*

Cozzolino, A., & Vakharia, A. J. (2007). 4PL's as "Orchestrators" of Logistics Networks. *Proceedings of the POMS Annual Meeting.*

De Angelis, R., Howard, M., & Miemczyk, J. (2018). Supply chain management and the circular economy: Towards the circular supply chain. *Production Planning and Control*, *29*(6), 425–437. doi:10.1080/09537287.2018.1449244

De Giovanni, P. (2016). State-and control-dependent incentives in a closed-loop supply chain with dynamic returns. *Dynamic Games and Applications*, *6*(1), 20–54. doi:10.100713235-015-0142-6

De Giovanni, P., & Zaccour, G. (2014). A two-period game of a closed-loop supply chain. *European Journal of Operational Research*, *232*(1), 22–40. doi:10.1016/j.ejor.2013.06.032

De Giovanni, P., & Zaccour, G. (2019). Optimal quality improvements and pricing strategies with active and passive product returns. *Omega*, *88*, 248–262. doi:10.1016/j.omega.2018.09.007

Ebeling, C. W. (1990). *Integrated Packaging Systems for Transportation and Distribution*. Marcel Dekker.

Eisenhardt, K. M. (1989). Building theories from case study research. *Academy of Management Review*, *14*(4), 532–550. doi:10.2307/258557

Eisenhardt, K. M., & Graebner, M. E. (2007). Theory building from cases: Opportunities and challenges. *Academy of Management Journal*, *50*(1), 25–32. doi:10.5465/amj.2007.24160888

Elia, V., Gnoni, M. G., & Tornese, F. (2017). Measuring circular economy strategies through index methods: A critical analysis. *Journal of Cleaner Production*, *142*, 2741–2751. doi:10.1016/j.jclepro.2016.10.196

Elkington, J. (2018). 25 years ago I coined the phrase "triple bottom line." Here's why it's time to rethink it. *Harvard Business Review*, *25*, 2–5.

Ellram, L. M. (1996). The use of the case study method in logistics research. *Journal of Business Logistics*, *17*(2), 93.

Farooque, M., Zhang, A., Thürer, M., Qu, T., & Huisingh, D. (2019). Circular supply chain management: A definition and structured literature review. *Journal of Cleaner Production*, *228*, 882–900. doi:10.1016/j.jclepro.2019.04.303

Fitzpatrick, L., Lewis, H., & Verghese, K. (2012). Implementing the Strategy. In *Packaging for Sustainability* (pp. 285–328). Springer. doi:10.1007/978-0-85729-988-8_8

Friedman, W. F., & Kipnees, J. J. (1977). *Distribution Packaging*. Krieger.

Frota Neto, J. Q., Walther, G., & Bloemhof, J. (2010). From closed-loop to sustainable supply chains: The WEEE case. *International Journal of Production Research*, *48*(15-16), 4463–4481. doi:10.1080/00207540902906151

García-Arca, J., Prado-Prado, J. C., & Garrido, A. T. G. P. (2014). "Packaging logistics": Promoting sustainable efficiency in supply chains. *International Journal of Physical Distribution & Logistics Management*, *44*(4), 325–346. doi:10.1108/IJPDLM-05-2013-0112

García-Arca, J., & Prado-Prado, J. C. P. (2008). Packaging design model from a supply chain approach. *Supply Chain Management*, *13*(5), 375–380. doi:10.1108/13598540810894960

Gawer, A., & Cusumano, M. A. (2014). Industry platforms and ecosystem innovation. *Journal of Product Innovation Management*, *31*(3), 417–433. doi:10.1111/jpim.12105

Geissdoerfer, M., Savaget, P., Bocken, N. M., & Hultink, E. J. (2017). The Circular Economy–A new sustainability paradigm? *Journal of Cleaner Production*, *143*, 757–768. doi:10.1016/j.jclepro.2016.12.048

Genc, T. S., & De Giovanni, P. (2020). Closed-loop supply chain games with innovation-led lean programs and sustainability. *International Journal of Production Economics*, *219*, 440–456. doi:10.1016/j.ijpe.2018.05.026

Genovese, A., Acquaye, A. A., Figueroa, A., & Koh, S. L. (2017). Sustainable supply chain management and the transition towards a circular economy: Evidence and some applications. *Omega*, *66*, 344–357. doi:10.1016/j.omega.2015.05.015

Ghisellini, P., Cialani, C., & Ulgiati, S. (2016). A review on circular economy: The expected transition to a balanced interplay of environmental and economic systems. *Journal of Cleaner Production*, *114*, 11–32. doi:10.1016/j.jclepro.2015.09.007

Gnoni, M.G., Tornese, F., Thorn, B.K., Carrano, A.L. & Pazour, J.A. (2018). *A measurement tool for circular economy practices: A case study in pallet supply chains*. Academic Press.

Govindan, K., & Hasanagic, M. (2018). A systematic review on drivers, barriers, and practices towards circular economy: A supply chain perspective. *International Journal of Production Research*, *56*(1-2), 278–311. doi:10.1080/00207543.2017.1402141

Govindan, K., Soleimani, H., & Kannan, D. (2015). Reverse logistics and closed-loop supply chain: A comprehensive review to explore the future. *European Journal of Operational Research*, *240*(3), 603–626. doi:10.1016/j.ejor.2014.07.012

Guide, V. D. R. Jr, & Van Wassenhove, L. N. (2006). Closed-loop supply chains: An introduction to the feature issue (part 1). *Production and Operations Management*, *15*(3), 345–350. doi:10.1111/j.1937-5956.2006.tb00249.x

Guo, S., Shen, B., Choi, T. M., & Jung, S. (2017). A review on supply chain contracts in reverse logistics: Supply chain structures and channel leaderships. *Journal of Cleaner Production*, *144*, 387–402. doi:10.1016/j.jclepro.2016.12.112

Harkham, A. (1989). *Packaging Strategy. CRC Press.*

Hellström, D., & Saghir, M. (2007). Packaging and logistics interactions in retail supply chains. *Packaging Technology and Science: An International Journal*, *20*(3), 197–216. doi:10.1002/pts.754

Howard, M., Hopkinson, P., & Miemczyk, J. (2019). The regenerative supply chain: A framework for developing circular economy indicators. *International Journal of Production Research*, *57*(23), 7300–7318. doi:10.1080/00207543.2018.1524166

Jacobides, M. G., Cennamo, C., & Gawer, A. (2018). Towards a theory of ecosystems. *Strategic Management Journal*, *39*(8), 2255–2276. doi:10.1002mj.2904

Jalali, H., Ansaripoor, A., Ramani, V., & De Giovanni, P. (2022). Closed-loop supply chain models with coopetition options. *International Journal of Production Research*, *60*(10), 3078–3106. doi:10.1080/00207543.2021.1910871

Kirchherr, J., Reike, D., & Hekkert, M. (2017). Conceptualizing the circular economy: An analysis of 114 definitions. *Resources, Conservation and Recycling*, *127*, 221–232. doi:10.1016/j.resconrec.2017.09.005

Kirwan, M. J., & Coles, R. (2011). *Food and beverage packaging technology*. John Wiley & Sons.

Kozik, N. 2020. Sustainable packaging as a tool for global sustainable development. In *SHS Web of Conferences* (Vol. 74, p. 04012). EDP Sciences. 10.1051hsconf/20207404012

Lambert, D., Stock, J. R., & Ellram, L. M. (1998). *Fundamentals of logistics management*. McGraw-Hill/Irwin.

Lindh, H., Williams, H., Olsson, A., & Wikström, F. (2016). Elucidating the indirect contributions of packaging to sustainable development: A terminology of packaging functions and features. *Packaging Technology & Science*, *29*(4-5), 225–246. doi:10.1002/pts.2197

Liu, J., Feng, Y., Zhu, Q., & Sarkis, J. (2018). Green supply chain management and the circular economy: Reviewing theory for advancement of both fields. *International Journal of Physical Distribution & Logistics Management*, *48*(8), 794–817. doi:10.1108/IJPDLM-01-2017-0049

Lockamy, A. III. (1995). A conceptual framework for assessing strategic packaging decisions. *International Journal of Logistics Management*, *6*(1), 51–60. doi:10.1108/09574099510805251

Massaroni, E., & Cozzolino, A. (2012). *Modelli della produzione industriale*. CEDAM.

Massaroni E. & Cozzolino A. (2021). Innovazioni logistiche per un packaging sostenibile: aspetti teorici ed evidenze empiriche. *Rivista di Studi Manageriali – Risma*, *1*(1), 109-128.

Massaroni, E., Cozzolino, A., & Wankowicz, E. (2015). Sustainability in supply chain management-a literature review. *Sinergie Italian Journal of Management*, *33*(Sep-Dec), 331–355.

Massaroni, E., Cozzolino, A., & Wankowicz, E. (2016). Sustainability reporting of logistics service providers in Europe. *International Journal of Environment and Health*, *8*(1), 38–58. doi:10.1504/IJENVH.2016.077662

Meherishi, L., Narayana, S. A., & Ranjani, K. S. (2019). Sustainable packaging for supply chain management in the circular economy: A review. *Journal of Cleaner Production*, *237*, 117582. doi:10.1016/j.jclepro.2019.07.057

Moula, M. M. E., Sorvari, J., & Oinas, P. (2017). Introduction: How can circular economy help to construct a more sustainable green circular society? In Constructing A Green Circular Society (pp. 1-5). University of Helsinki.

Nasir, M. H. A., Genovese, A., Acquaye, A. A., Koh, S. C. L., & Yamoah, F. (2017). Comparing linear and circular supply chains: A case study from the construction industry. *International Journal of Production Economics*, *183*, 443–457. doi:10.1016/j.ijpe.2016.06.008

Paine, F. A. (Ed.). (1991). *The packaging user's handbook*. Springer Science & Business Media.

Potting, J., Hekkert, M. P., Worrell, E., & Hanemaaijer, A. (2017). Circular economy: Measuring innovation in the product chain (No. 2544). PBL Publishers.

Preuss, L. (2009). Addressing sustainable development through public procurement: The case of local government. *Supply Chain Management*, *14*(3), 213–223. doi:10.1108/13598540910954557

Ramani, V., & De Giovanni, P. (2017). A two-period model of product cannibalization in an atypical Closed-loop Supply Chain with endogenous returns: The case of DellReconnect. *European Journal of Operational Research*, *262*(3), 1009–1027. doi:10.1016/j.ejor.2017.03.080

Rezapour, S., Farahani, R. Z., Dullaert, W., & De Borger, B. (2014). Designing a new supply chain for competition against an existing supply chain. *Transportation Research Part E, Logistics and Transportation Review*, *67*, 124–140. doi:10.1016/j.tre.2014.04.005

Richardson, M. (2011). Design for reuse: Integrating upcycling into industrial design practice. In *International Conference on Remanufacturing* (pp. 1-13). Academic Press.

Robertson, G. L. (2006). *Food Packaging: Principles and Practice*. CRC Press.

Saghir, M. (2002). Packaging information needed for evaluation in the supply chain: The case of the Swedish grocery retail industry. *Packaging Technology and Science: An International Journal, 15*(1), 37–46. doi:10.1002/pts.565

Saghir, M. (2004). The concept of packaging logistics. In *Proceedings of the Fifteenth Annual POMS Conference* (pp. 1-31). Academic Press.

Saghir, M., & Jönson, G. (2001). Packaging handling evaluation methods in the grocery retail industry. *Packaging Technology and Science: An International Journal, 14*(1), 21–29. doi:10.1002/pts.523

Shapiro, R. D., & Heskett, J. L. (1985). *Logistics Strategy: cases and concepts*. West Group.

Silva, N., & Pålsson, H. (2022). Industrial packaging and its impact on sustainability and circular economy: A systematic literature review. *Journal of Cleaner Production, 333*, 130165. doi:10.1016/j.jclepro.2021.130165

Sonneveld, K., James, K., Fitzpatrick, L., & Lewis, H. (2005). Sustainable packaging: how do we define and measure it. In *22nd IAPRI Symposium* (pp. 1-9). Academic Press.

Soroka, W. (2002). *Fundamentals of Packaging Technology*. Institute of Packaging Professionals.

Souza, G. C. (2013). Closed-loop supply chains: A critical review, and future research. *Decision Sciences, 44*(1), 7–38. doi:10.1111/j.1540-5915.2012.00394.x

Stake, R. E. (2005). Qualitative case. The SAGE Handbook of Qualitative Research, 1, 443-466.

Sung, K. (2015). April. A review on upcycling: Current body of literature, knowledge gaps and a way forward. *World Academy of Science, Engineering and Technology*.

Sung, K., Singh, J., & Bridgens, B. (2021). *State-of-the-Art Upcycling Research and Practice*. Springer International Publishing. doi:10.1007/978-3-030-72640-9

Tornese, F., Gnoni, M. G., Thorn, B. K., Carrano, A. L., & Pazour, J. A. (2021). Management and Logistics of Returnable Transport Items: A Review Analysis on the Pallet Supply Chain. *Sustainability, 13*(22), 12747. doi:10.3390u132212747

Twede, D. (1992). The process of logistical packaging innovation. *Journal of Business Logistics, 13*(1), 69.

Twede, D., & Parsons, B. (1997). *Distribution packaging for logistical systems: A literature review*. Pira.

Verghese, K., & Lewis, H. (2007). Environmental innovation in industrial packaging: A supply chain approach. *International Journal of Production Research, 45*(18-19), 4381–4401. doi:10.1080/00207540701450211

Vernuccio, M., Cozzolino, A., & Michelini, L. (2010). An exploratory study of marketing, logistics, and ethics in packaging innovation. *European Journal of Innovation Management, 13*(3), 333–354. doi:10.1108/14601061011060157

Weetman, C. (2017). *A circular economy handbook for business and supply chains: Repair, remake, redesign, rethink*. Kogan Page Publishers.

Wegener, C. (2016). Upcycling. In *Creativity—A New Vocabulary* (pp. 181–188). Palgrave Macmillan.

Williams, H., Wikström, F., & Löfgren, M. (2008). A life cycle perspective on environmental effects of customer focused packaging development. *Journal of Cleaner Production, 16*(7), 853–859. doi:10.1016/j.jclepro.2007.05.006

Yam, K. L. (2009). *The Wiley Encyclopedia of Packaging Technology*. Wiley. doi:10.1002/9780470541395

Yin, R. K. (2003). Applied social research methods series: Vol. 5. *Case study research design and methods* (3rd ed.). SAGE.

Ying, J., & Li-jun, Z. (2012). Study on green supply chain management based on circular economy. *Physics Procedia, 25*, 1682–1688. doi:10.1016/j.phpro.2012.03.295

ENDNOTES

1 https://ec.europa.eu/environment/pdf/circular-economy/new_circular_economy_action_plan.pdf (accessed 18 April 2022)

2 https://eur-lex.europa.eu/legal-content/EN/TXT/?qid=1583933814386&uri=COM:2020:98:FIN (accessed 18 April 2022)

3 https://ellenmacarthurfoundation.org/topics/circular-economy-introduction/overview (accessed 18 April 2022)

4 https://archive.ellenmacarthurfoundation.org/explore/the-circular-economy-in-detail (accessed 18 April 2022)

5 https://eur-lex.europa.eu/legal-content/EN/TXT/?qid=1583933814386&uri=COM:2020:98:FIN (accessed 18 April 2022)

6 https://sustainablepackaging.org (accessed 18 April 2022)

7 http://www.europen-packaging.eu (accessed 18 April 2022)

8 https://www.echolabs.fercam.com/it

9 The analysis is based on desk research from the institutional website of Fercam Echo Labs (https://www.echolabs.fercam.com/it), LinkedIn news, and official documents. Part of text and contents of this Section is also from the institutional website of Fercam (https://www.fercam.com/en/welcome-1.html).

10 https://www.euromerci.it/in-primo-piano/presentate-a-roma-fercam-green-logistics.html (accessed 18 April 2022)

11 https://www.statista.com/statistics/1228432/global-pallet-market-size/ (accessed 18 April 2022)

12 https://www.statista.com/statistics/1228432/global-pallet-market-size/ (accessed 18 April 2022)

13 https://www.rilegno.org/wp-content/uploads/2017/12/La-filiera-degli-imballaggi-di-legno.pdf (accessed 18 April 2022)

14 https://www.rilegno.org/ (accessed 18 April 2022)

15 https://ellenmacarthurfoundation.org/topics/circular-economy-introduction/overview (accessed 18 April 2022)

Chapter 2
Battery Swapping Business Model:
The Case of Lithion Power

Nandan Kumar Singh
https://orcid.org/0000-0003-1963-9114
Indian Institute of Management, Visakhapatnam, India

Vinay Ramani
Indian Institute of Technology, Kanpur, India

EXECUTIVE SUMMARY

Lithion Power is India's leading provider of battery as a service, supplying lithium-ion batteries for e-bikes and three-wheelers. They use the battery swapping technology to power electric vehicles. Lithion's omni-charging technology allows electric vehicles to either charge at home, at public chargers or at a Lithion Swapping Point (LSP), primarily spread across North Delhi and the nearby state of Haryana. At these LSPs, a customer can drive in with their existing battery and in a less-than-five-minute walk away with a charged battery. Lithion is collaborating with major operators and original equipment manufacturers (OEMs) to make an all-electric India a reality by 2030. Key to Lithion's business model is a battery management system (BMS) that allows the optimal use of the battery, thereby extending its use life. This chapter discusses Lithion's battery swapping service model and puts it in the context of circular economy.

DOI: 10.4018/978-1-6684-5001-7.ch002

ORGANIZATION BACKGROUND

Piyush Gupta, an Indian Institute of Technology, Kharagpur graduate with an MBA from INSEAD, founded Lithion Power in 2017 with Chandrashekhar Bhide, a computer science graduate from the Indian Institute of Technology, Bombay with MBA from the Indian Institute of Management, Ahmedabad. Lithion originally began with the idea of developing an electric bus, which was a genuinely audacious endeavor. After a year of learning, they shifted their focus to batteries, which resulted in the formation of Lithion Power. It began with an agenda devised by India's most renowned and technically competent battery specialists. They initially focused on developing a Battery Management System (BMS) tailored to Indian conditions and a telemetry device before launching a battery/energy as a service business. Lithion began commercial operations in 2018. Currently, it has over 250 assets on the platform. Lithion focuses on the business-to-business (B2B) industry. It has a presence in Delhi, India and the National Capital Region (NCR), as most EVs are based in and around Delhi and NCR.

The Indian government is taking multiple initiatives to boost EV adoption in the country, including the implementation of different subsidies and incentives for manufacturers and automobile purchasers. The benefits vary according to the Indian state in which the consumer resides. State governments have subsidized EV adoption by offering incentives to electric car customers in addition to the federal government's subsidies. Delhi is the state in India that provides the most significant incentives for electric vehicles (CarDekho, 2021). It is equivalent to the federal government's subsidies. As a result, the benefits of purchasing an electric vehicle in Delhi can be roughly double of those in other states. Additionally, under the existing program, the state government waives registration and road tax on all electric vehicles. The Delhi government website (Delhi, n.d.). presently lists 169 charging stations and 377 charging points, making this one of the largest cities in the country to have such an infrastructure.

Lithion was a finalist in the Electricite de France (EDF) Innovative products and services that improve the value chain of e-mobility infrastructure, including charging and battery swapping stations, EV system components such as the electricity network, energy production, and storage, and was also named to the Economic Times' list of India's Top Innovative MSMEs (Economic Times, 2021) because of their *omni-charging technology* which allows electric vehicles to either charge at home, at public charging points, or at Lithion's swap systems. This article first describes Lithion's business model in the EV ecosystem and thereafter draws a connection of Lithion's business model to the notion of Circular Economy.

Lithion is now active in the Delhi and NCR region, with plans to expand to other cities. In Bengaluru, it is in the process of establishing similar stations. In the next few years, it plans to open at least 500 battery swap facilities in Delhi and 10,000 across the country. It currently employs over 50 people and has over 250 registered end users on its platform.

SETTING THE STAGE: LITHION'S BUSINESS MODEL

Lithion's core business model is to offer Battery Swapping as a Service. Battery swapping or battery swapping as a service allows a user to replace their existing discharged battery with a fresh charged one at a swapping station. Lithion does not manufacture the battery, and also does not own the swapping stations. It deploys the batteries at the station using a franchise model to keep the costs minimal. The franchise locations too are chosen very carefully - they are connected to the EV ecosystem. They are either dealers of vehicles, dealer of batteries or EV accessories. So the franchise have a commercial set up. Moreover, the space required for charging the batteries is very minimal. They need a 25 sq.ft. space because only the battery is being charged not the entire vehicle. At the centre of its business model innovation is the Battery Management System (BMS). We elaborate on the BMS below.

Lithion Battery Management System

On Lithion's platform which is used by the Battery Swapping Station, both the vehicle and the battery are registered. The battery pack includes a smart component system which is the heart and soul of the battery and is responsible for the effective and efficient BMS. Each of these battery packs is equipped with Global Positioning System (GPS) and Global System for Mobile Communications (GSM) capabilities. They perform real-time monitoring and tracking of 35-40 parameters of the battery, including individual cell voltage, current drawn, remaining capacity, the temperature at various places inside the battery pack, vehicle coordinates, and certain driver-related and vehicle-related information. Data is sent to the cloud every 1-2 seconds, and the battery package unit includes local storage for data backup in the event of a network outage. The BMS syncs the data on the cloud by sending the local data at a faster rate of 2X or 4X once the network is restored. The battery is monitored in real-time to adjust critical parameters for best performance in various settings, such as when a vehicle operates on a slope, summer, or rainy weather.

Lithion's primary customers are three-wheelers and two-wheelers (electric bikes and scooters). These vehicles are very popular with e-commerce delivery agencies. Typically e-commerce vehicles will travel approximately 60-100 kilometers a day. Sometimes the same vehicle will be used by two different drivers who work in shifts. Take for example, the milk delivery service operation which starts at early morning hours at 5 a.m. and continues until approximately 9 a.m. Thereafter, the vehicle is used by the second driver for other e-commerce delivery. As a result, a downtime charging of 4-6 hours is not feasible during the day. The BMS complements the battery swapping service by ensuring that everything is geared towards maximizing utilization of the deployed assets because such an objective is directly proportional to the revenue generated.

In a 3-5 km range, the BMS tells the driver the specific locations where they can go and get their batteries swapped. If there are 4 such swapping stations and only 3 of them have batteries which can be given out in the next 15 minutes, then only those will be shown to the driver. Furthermore, the BMS has historical data to predict the usage based on the past 30-60-90 days travel distance. Once the driver reaches the battery swapping station, they just mention their mobile number which is their unique identification number for the battery. Lithion has designed a mobile application, the details of which are given in Figure 1 below. The Lithion Mobile Application has four specific features: ***(1)****: Real Time Availability and Live Location,* ***(2)****: Real Time State of Affair of Battery,* ***(3)****: Booking and Payment, and* ***(4)****: Notification.* The details of the four features are also provided in Figure 1.

(1) ***Real Time Availability and Live Location*****:** This feature helps the driver or the fleet operator to find a nearby swapping station and assists them in reaching to that station. It will also provide the information about how many batteries are available for swapping in all the nearby swapping stations. This feature is important because connecting the driver to the nearest battery swapping station not only is efficient, it saves substantial cost for the user and in addition also, manages the inventory of electric vehicles effectively.

Figure 1. Lithion Mobile Application Feature

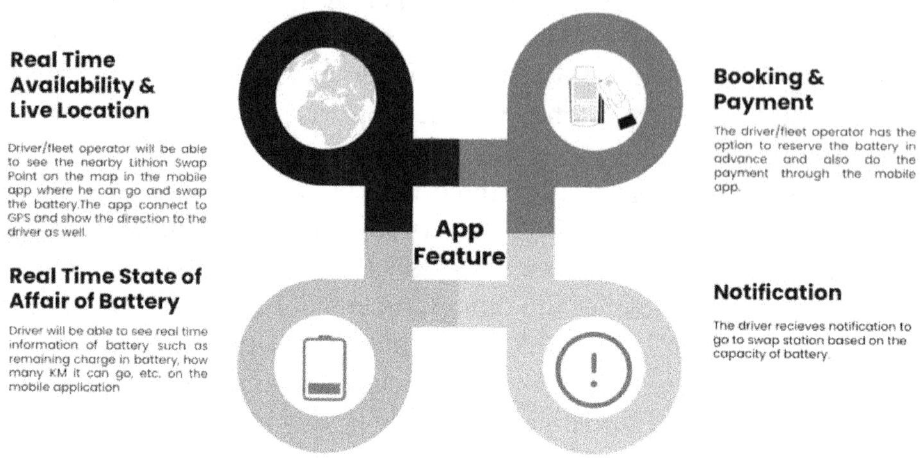

(2) ***Real Time State of Affair of Battery*:** This feature helps the driver to know the current status of the battery. The mobile application shows the charging left, how much more KM vehicle can go etc. This will provide a real time information about the battery which will help them to take decision about when to go for swapping. Since the real time data is captured by Lithion Power, this feature helps in better management of the health of the battery and also helps in prolonging its life.

(3) ***Booking and Payment*:** The drivers have the option to book the battery in advance. The Battery Management System provides several conveniences for payment. For example, the driver or fleet operator can make the payment after swapping the battery at the swapping station, or, can pay while booking as well. The Battery Management System also allows convenience in making the payment through mobile phones as well as other payment application platforms.

(4) ***Notification:*** The advance feature sends the notification/warning to driver about the battery's remaining charge and directs them to the nearby swapping station to avoid any situation where battery is drained out completely. This feature has many interesting aspects. A completely drained battery requires a longer time to be fully charged, so, informing the driver earlier saves significant time and use of electricity for charging.

At the battery swapping station, the intelligent BMS system automatically identifies the battery for swap during swapping. It displays the battery location

Figure 2. Different Phases of Customer's Journey & Circular Economy Point

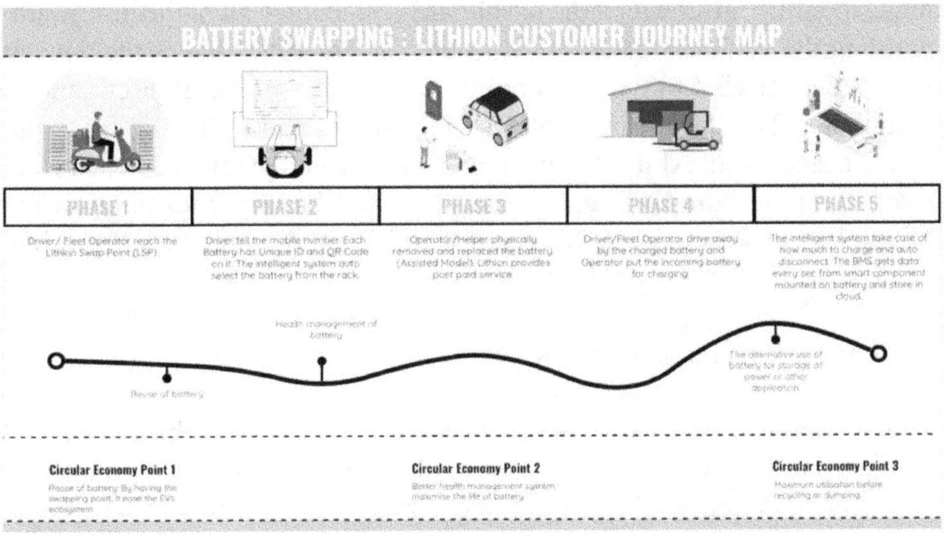

(row and column of the rack) on the dashboard to swap a particular vehicle battery based on the driver and vehicle information. The BMS suggests the battery (say a range of 80 kms) instead of 60 kms or 100 kms.

The battery weighs approximately 20 kgs for a 3-wheeler and about 6 kgs for a 2-wheeler and there is a connector which plugs in the new battery. After changing the batteries, the discharged battery is placed on the rack. The operator at the swap station removes the charged battery meant for use in the vehicle and then removes the used battery and places it on the vacant slots meant for charging. The BMS automatically indicates the level to which the discharged battery needs to be charged and the intelligence system disconnects it when that level is reached.

Figure 2. depicts the various phases of a customer's journey. To summarize, there are essentially five connected phases:

Phase 1: *Registration of the vehicle and battery on the Lithion platform.*

The driver or fleet operator registers their vehicle with the Lithion. Their details is fed into the system and updated on the Lithion platform.

Phase 2: *Service at Lithium Swap Point (LSP).*

The driver provides his or her mobile phone number (which serves as a unique identity), after which the dashboard will display all of the information regarding the vehicle's battery. It displays information such as the date and time of the last visit to the swap point, the amount of energy used, and the amount due. The dashboard displays the battery for the specific driver and vehicle based on an AI model that considers various parameters such as the driver's driving competence, service area, and battery consumption rate. The intelligent system picks the battery which is best suited for the vehicle and driver.

Phase 3: *Assisted mode of service.*

The operator removes the vehicle's battery and replaces it with the designated battery from the rack. As it is a post-paid service, the driver is charged for the last use of energy. Thus, it is important to note that the driver pays for the energy that the vehicle has consumed.

Phase 4: *Post replacement service.*

The operator places the drained battery on the rack for charging while the driver makes the payment and drives away with the charged battery. The Battery Management System indicates to the operator at the charging station in which zone the battery needs to be placed at.

Phase 5: *Charging Process.*

The discharged battery is placed to the charging racks by the operator. The smart BMS system decides how much to charge automatically. When the charging level reaches that point, the battery automatically disengages from the charger.

The current battery swapping service model is an assisted model and not a self-service model. Once the operator at the service station enters the mobile number on the dashboard, they come to know about all the details regarding the battery, and in particular about the energy left on a real time basis. For the convenience of the driver the range is specified. The vehicle operator pays for the energy that the vehicle has utilized (equivalently, the distance driven). Formally, the payment is

$$Payment = (E_0 - E_1)r$$

Figure 3. Lithion Power Supply Chain

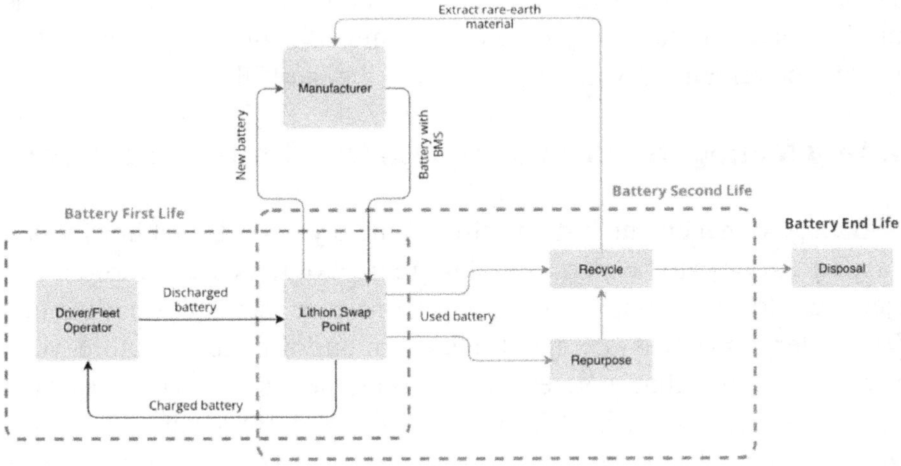

where, denotes the energy at the time the battery was provided, denotes the energy at the time of replacement of the battery, and *r* denotes rate per unit of energy used. The BMS also allows the payment to be made digitally by connecting the user to the various payment platforms. Interestingly, in contrast to vehicles which involve refueling with petrol/diesel where the payment is prepaid, Lithion follows a *postpaid* model. The primary objective is to secure the assets (in this case, the battery).

The postpaid system makes business sense for a fleet operator serving EVs without the battery as the upfront cost of the vehicle goes down. Further, as is the case with a petrol or diesel vehicle, one does not pay for a lifetime of energy upfront. A customer pays as you go. Figure 3. presents the schematic of the supply chain. We discuss the aspects of the first stage and second stage in the next section.

Lithion Power Supply Chain

The driver registered the vehicle with the Lithion. The driver swapped its discharged battery with the charged battery at the Lithion Battery Swapping Station. Lithion puts the discharged battery for the charging at the swapping station. The details is updated in the Lithion platform and the mobile application. After the first life of battery, Lithion decides whether to use those battery for repurpose or sell it to other entity who will repurpose it or to the entity who will recycle it. When the battery is repurposed, after the second life it will be recycle and the rare earth material will be extracted from it which will be used by the battery manufacturer to produce the new battery. The remaining will be disposed to the landfill.

Battery Management System and the Circular Economy

The Battery Management System (BMS) is the key towards enabling the battery swapping services to be an essential ingredient of the Circular Economy. Its objective is to ensure optimal battery performance thereby extending the life of the battery as well as keeping it safe - by monitoring temperature, voltage, current flow *etc.*; that ensures that the parameters are within a standard operating range. The World Economic Forum defines Circular Economy (World Economic Forum, n.d.) as "*an industrial system that is restorative or regenerative by intention and design. It replaces the end-of-life concept with restoration, shifts towards the use of renewable energy, eliminates the use of toxic chemicals, which impair reuse and return to the biosphere, and aims for the elimination of waste through the superior design of materials, products, systems and business models.*"

To understand how Lithion is contributing to the Circular Economy, we refer the reader to Figure 1 and Figure 4. An EV battery has primarily two phases, called as first life and second life. In the first life of the lithium-ion battery, there is a degradation of the battery in the first five years of its service. At the end of the first life, manufacturers face three distinct options - *(i)* reuse the battery, thus giving it a second life, *(ii)* recycle, which is economically feasible as the battery electrodes contain rare earth metals such as cobalt and nickel, and *(iii)* disposal which occurs if the battery pack is damaged. Reuse occurs when there is a demand or market for energy storage for applications which are not very energy intensive. On the other hand, most regulations prevent disposal of batteries in landfills.

As explained in Figure 3., on the Lithion platform, the driver or fleet owner registers their battery. Lithion Power provides the driver with a charged battery. It sends the new battery (registered as a new vehicle battery) to the manufacturer, who installs the innovative component (designed by Lithion) within the battery pack, which serves as the intelligence of the battery management system. Lithion receives the batteries from the manufacturer and charges them at several swap stations.

Lithion contributes to the circular economy at three points:

Circular Economy Point 1: Battery Swapping as a Business Model (Reuse of battery). It will have maximum utilization of the first life of the battery. This helps in prolonging the use of the product, in this case the electric battery.

Circular Economy Point 2: Smart BMS increase the first life of battery by ensuring that operation is within the standard operating range. The smart BMS tracks and monitors the State of Health (SOH) of battery through various parameters like current, voltage, temperature at different point in battery pack etc. Complemented with Lithion's BMS, optimal performance of the battery is thereby ensured.

Circular Economy Point 3: Smart BMS also increases the second life of the battery and thus recycler can use it for storage purpose (Repurpose of battery after first life). It will have maximum utilization of the second life of the battery. This aspect is also important as secondary use of the battery is only attractive when there are significant economic gains associated with that use. Ensuring good health of the battery is an important first step towards achieving this goal.

One way in which Lithion is directly contributing to a circular economy is by the use of a smart BMS which extends the life of the battery, and also reduces the wear and tear of the battery. This extends the first life of the battery directly. Secondly, batteries entering the second life or reuse have a gradual degradation that happens over its lifetime. Batteries which have finished their first life are deployed in energy storage devices in hotels, dormitories, and hostels, where they provide back up. Since a Battery on day one of year one is different from the battery on day one of year two, which in turn is different from that on day one of year three, and so on, the smart BMS helps Lithion identify the optimal usage and the transition of the battery from the first life to the second life.

The BMS also helps the recyclers to understand the steady state of affairs of their batteries. The recycling ecosystem is still in its infancy and will take another two to three years for it to fully develop as lithium-ion batteries are a recent phenomenon. In lithium-ion batteries, we cannot recover more than 7-8% of lithium which is a light metal and hence it is in everyone's interest try to repurpose or reuse the battery for other applications. For example, auto manufacturer, Nissan has formalized a partnership with Sumitomo Corporation to reuse battery packs from the Nissan Leaf for stationary distributed and utility-scale storage systems (Hauke E, Patrick H, and Giulia S, 2019).

LITERATURE REVIEW

As environmental degradation is putting an increasing pressure on our planet, the conventional linear economic system of ``take-make-dispose'' is no longer sustainable. In a study conducted by Kaza et al. (2018) and as reported by Maranesi and De Giovanni (2020), the annual waste generation is estimated to have an increase of 70% by 2050. Increased material consumption and unsustainable extraction of resources are primary causes of this. In this context, there has been an increased attention paid by academicians, practitioners, and policy makers to the concept of circular economy. Kirchherr et al., (2017) define circular economy as ``an economic system that is based on business models that replace the 'end-of-life' concept with reducing, alternatively reusing, recycling and recovering materials in production/distribution and consumption processes; thus, it applies at the micro level (products, companies, consumers), meso level (eco-industrial parks), and macro level (city, region, nation and beyond), with the aim to accomplish sustainable development by creating environmental quality, economic prosperity, and social equity, and benefits for both the current and the future generations''.

Circular Economy is becoming a critical part of any firms corporate strategy (D'Amato et al., (2017), Landrum and Ohsowski (2018), Maranesi and De Giovanni (2020)). Firms are modifying their corporate strategies by integrating sustainability into their long-term goals (see De Giovanni (2022) and De Giovanni and Folegiero (2022)). An important strategy in this regard is the shift from a waste-stream to a value-stream approach (De Giovanni (2021)). As mentioned by Maranesi and De Giovanni (2020), the strategies that firms can adopt include prolonging the lifespan of the product, reuse and redistribute the product to multiple users, remanufacture without disassembling the product, recycle to reduce the use of virgin materials, and cascades to implement the biological cycle till the product naturally degrades. Furthermore, the authors present case studies of four Italian firms engaged in circular economy and conclude that, circular economy is the "true business lever'' that achieves social, environmental, and economic performance much beyond the minimum. Our case study on Lithion Power complements the existing literature by focusing on a particular service strategy to prolong the life span, reuse and redistribute, remanufacture and recycle.

One of the primary requirements of circular economy in the case of electric batteries is that the residual power and valuable metal components are used to their maximum potential by the time the vehicle reaches its end of first life.

Thereafter, remanufacturing and recycling occurs at various stages after which the battery becomes redundant. Remanufacturing (or reusing) and recycling are crucial to the success of circular economy in the Electric Vehicles ecosystem. However, significant challenges remain in recycling electric batteries (Ahuja et al., (2020)). At the end of their first life, electric vehicles may still contain enough residual capacity for secondary use (Ahmadi et al., (2014)). Recycling to recover rare earth and critical metals like cobalt, lithium, nickel etc., has to be undertaken in ways that do not impose significant cost to the environment (Hirst et al., (2018), Hill et al., (2019)). This is because lithion ion batteries contain many hazardous substances which may pose a risk both to human life as well as the environment. Furthermore, proper regulatory frameworks need to be designed. One framework as suggested by Ahuja et al., (2020) is the servitisation model, where the manufacturers retain ownership of the assets and the customer pays for the service. In this regard, our case complements the servitisation model where the consumer does not own the battery and rather pays for the energy consumed.

Adoption of a circular economy framework requires a fundamental shift in the business-as-usual approach of firms. It involves reduction of waste and pollution, reuse of materials, and regenerating natural systems (see, Maranesi and De Giovanni (2018)). Dobrotă and Dobrotă (2018) argue that implementation of a circular economy framework requires a synthesis of the following business models – (i) Circular inputs involving the use of completely recyclable materials and the use of renewable energy to reduce environmental footprint, (ii) Sharing Platforms involving collaboration between the consumers and the service providers for access to goods and services, (iii) Product as a service concept involving leaving the ownership of the product to the manufacturer, (iv) Resource recovery involving usable energy from waste products, and (v) Product use extension involving increasing the life cycle of the product through technological advancements, upgrading and resale. While implementation of all the above features in the context of an electric vehicle ecosystem is challenging, considerable progress has been made on some of the features. For example, a Battery Management System is an technological advancement in the EV ecosystem which prolongs the lifespan of the electric battery, as well as achieves the product as a service concept as it indicates effective use of energy that can be use in a secondary source after the battery completes its first life.

Challenges in Reuse and Recycling

Both reuse as well as recycling involve various challenges. In reuse, one primary consideration is that reuse is economically viable only if the cost differential between a new battery and a used battery is approximately 50 - 70%. However, with learning, the cost differential may decrease to a level of approximately 25%, at which point reuse is not an attractive model. On the technological end, for reuse to be effective, all the cells have to be at the same state of health-for example it cannot be that one cell is three years old, other is three months old and the third is a fresh one. Such a setting will destabilize the whole battery. Since the strength of a supply chain is its weakest link, the performance of the battery is dependent on the weakest cell in the battery pack. Therefore, Lithion during its tests ensures that all the cells operate at the approximately the same level. In India, battery pack manufacturers assemble cells from different sources and many don't use a cell testing machine as it is an expensive device.

The challenges related to recycling arise from the fact that no large player (or collector) is getting into recycling parts because there is no steady supply of degraded batteries. There is no player who will aggregate all of this across locations and bring it up to some central location. The central facility technology is also not well developed. To this extent Lithion's BMS may provide help to recyclers by helping them collect the batteries (which need recycling) from the battery swapping stations.

The Lithion BMS increase the first life as well as the second life of the battery. It also help to determine the optimal level of battery state of health for first life and second life i.e., M_1 and M_2. Figure 4 depicts this scenario.

M_1 is the level of health of the battery that determines when it enters its second life. The age (year/month) at which a battery attains the M_1 state of health is referred to as the battery's first life, *f*. M_2 is the condition of health of the battery that indicates when it is time to dispose it of in a landfill.

The age (year/month) at which a battery reaches its M_2 state of health is referred to as the battery's end life, i.e., *s*. The first and second lives of the battery are increased to *f* ' and *s'* respectively, using Lithion's in-house designed smart BMS. In Figure 3, the BMS improves the health of the battery as depicted by the dotted curve. Additionally, the BMS assists in determining the M_1 and M_2 levels, i.e., the optimal level at which the battery can be re-purposed for smaller batteries, storage/power back up in off-grid places, hotels, and hostels, and the level at which it is advisable to dispose it

Figure 4. Lithion Battery Management System Impact on Battery life

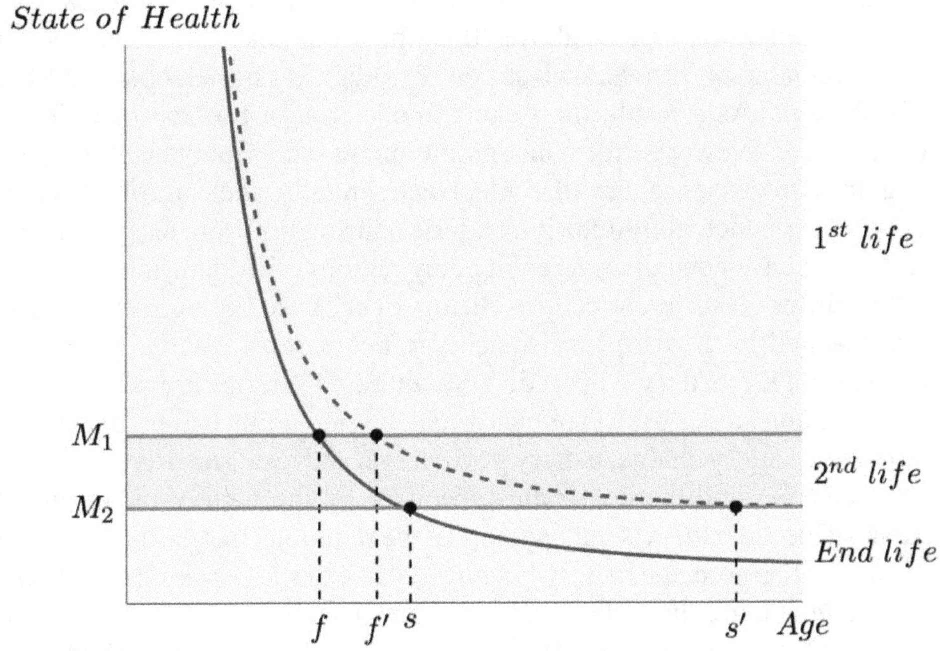

in landfill. Therefore, maximum battery use is possible in both the first and second life of the battery with the help of Lithion's intelligent BMS.

ROLE OF BATTERY MANAGEMENT SYSTEM IN A CIRCULAR ECONOMY

Batteries are complex systems. Capacity and internal resistance are the two battery properties that capture the most crucial battery performance characteristics. Capacity measures the total electric charge that a battery can deliver. Internal resistance is a figure-of-merit for the opposition to a current flow in a battery. Both capacity and resistance are heavily influenced by the operating and load conditions. Battery performance deteriorates through time and usage, resulting in a drop in capacity and an increase in resistance. The performance of a battery diminishes with time and use, manifesting as a decline in capacity and an increase in resistance. BOL performance is commonly described as 100 per cent state-of-health (SOH), a percentage that decreases for degrading battery performance (explained in figure 4).

A battery is often regarded to have reached its end-of-life for vehicle application when the battery's performance values has gone below 80% SOH or deterioration level increase above 20%. The battery can still be utilized for various secondary purposes, such as power storage in a hostel, hotel, or other establishments. As a result, the second application of traction batteries in applications with lower performance requirements represents a viable market.

Ageing processes cause the electrochemically measurable loss of battery performance in lithium-ion batteries during time and usage. Ageing mechanisms have been discovered in many regions of lithium-ion batteries, and they differ for different battery chemistries. These ageing mechanisms, which resulted in increased resistance and decreased capacity, have been dependent on the battery's operating conditions. Batteries are subjected to many operation modes used in onboard vehicles, including battery discharge, regenerative battery charge, battery grid charge, and rest. The driving events, driver behavior, and the constraints specified by the battery management system define the battery's duty cycle, i.e. the required charge and discharge current. Furthermore, the battery is subjected to external working conditions such as temperature, humidity, and vibration. The operating circumstances have a considerable impact on the performance of automobile batteries. The onboard utilization of battery systems must be examined to optimize the battery performance. Under controlled conditions, laboratory studies can provide essential insight into battery calendar and cycle life, exposing temperature, cycling depth, and SOC dependence. However, because the conditions during vehicle operation are complicated, as shown above, those tests have limited value for real-world applications. Because the battery is sensitive to its operating conditions, it must be managed to optimize its performance, dependability, and lifetime. As a result, battery management systems (BMSs) are used.

The major function of the BMS is to ensure the safe operation of the battery. This is accomplished by giving accurate readings of voltage (cell voltages), current, and temperature to comply with usage restrictions, i.e., safeguard the battery from damage caused by overheating, overcharging, and over-discharging. Ideally, battery management also ensures that the battery is used as efficiently as possible by maintaining it in optimal conditions, such as through cell balancing and a climate system. Battery status information is required to optimize battery consumption. The state estimation can determine how much power and energy are available for vehicle propulsion and other subsystems. Ultimately, optimizing battery utilization for the battery's SOH will result in longer battery life.

COMMENTARY ON BATTERY SWAPPING POLICY IN INDIA

The Battery Swapping Policy of India (National Institutional for Transforming India (NITI), 2022) would support the vision of catalyzing large-scale EV adoption by improving the efficient and effective use of rare resources by promoting battery swapping technology implemented through Battery as a service (BaaS) business models, ensuring lower upfront costs and minimal downtime and lower space requirements. The policy tackles critical technical, regulatory, institutional, and financial obstacles that will aid in the development of battery swapping ecosystems and enable large-scale battery swapping adoption in India.

The key objectives of this policy reproduced from the NITI Ayogya report (National Institutional for Transforming India (NITI), 2022) are as follow:

- *Promote swapping of batteries with Advanced Chemistry Cell (ACC) batteries to decouple battery costs from the upfront costs of purchasing EVs, thereby driving EV adoption.*
- *Offer flexibility to EV users by promoting the development of battery swapping as an alternative to charging facilities.*
- *Establish principles behind technical standards that would enable the interoperability of components within a battery swapping ecosystem, without hindering market-led innovation*
- *Leverage policy and regulatory levers to de-risk the battery swapping ecosystem, to unlock access to competitive financing.*
- *Encourage partnerships among battery providers, battery OEMs and other relevant partners such insurance/financing, thereby encouraging the formation of ecosystems capable of delivering integrated services to end users.*
- *Promote better lifecycle management of batteries, including maximizing the use of batteries during their usable lifetime, and end of life battery recycling.*

”

The policy specifies the minimum technical and operational characteristics that battery swapping ecosystems must meet in order to enable the implementation of battery-swapping infrastructure that is effective, efficient, reliable, safe, and user-friendly. It aims to encourage the use of electric vehicles (EVs) by lowering the upfront costs of purchasing an EV compared to the cost of an internal combustion engine (IEC) vehicle by providing direct and indirect financial incentive to battery providers and EV users. To stimulate private sector engagement and attract cheap financing, this policy stresses permitting innovation in the adoption of possible business models is essential infrastructure. Furthermore, it emphasizes the necessity of reusing end-of-first-life swappable batteries and recycling end-of-life batteries establishes the framework for developing unique battery codes for ACC batteries covered by it. Finally, it also establishes an institutional framework to aid in the on-the-ground execution of the essential battery-swapping infrastructure and achieve the stated policy goals. It was designed to encourage the usage of battery swapping systems in light electric power train vehicles (LEVs) and e-rickshaw/e-carts.

Battery Swapping Service Implementation

The deployment of battery swapping networks will necessitate coordination and support from state and municipal governments. States and/or local governments should ensure that battery swapping is treated equally, with plug-in charging available for public (and captive) Battery Charging Station (BCS) and Battery Swapping Station (BSS), as required.

The Bureau of Indian Standards (BIS) and other relevant organizations will draught regulations for minimum battery performance and durability requirements to address concerns about battery life.

BIS will create guidelines for reusing and re-purposing of end-of-first-life batteries from EV applications to ensure the safety, reusability, and long-term viability of business models in the second-life application of used EV batteries. The Bureau of Indian Standards (BIS) or the Ministry of Environment, Forest and Climate Change (MOEFC) will define the battery ownership and obligation to ensure proper end-of-life (EOL) recycling of EV batteries. The improper disposal of electric vehicle batteries in landfills or scrap will be prohibited. Separate Battery Management Rules will be published to address batteries' end-of-life handling in detail and establish the Extended Producer Responsibility (EPR).

The responsible organizations for the rollout of battery swapping stations are as follow:

The Bureau of Energy Efficiency (BEE) deploying EV public charging infrastructure, will oversee the nationwide implementation of battery swapping networks. States and union territories (UTs) are responsible for the implementation and administration of the ecosystem for battery swapping.

State nodal agencies (SNAs) for electric vehicle (EV) public charging infrastructure will assist in the deployment of battery swapping. The following state agencies will assist in support of SNAs:

- *Transport Departments and State Transport Authorities* are responsible for streamlining the registration process for vehicles sold without batteries or with battery charging capabilities.
- *Municipalities* are responsible for battery-swapping station planning, authorizations, and land allocation.
- *Energy departments and electricity distribution companies* are responsible for powering battery swapping stations and providing policy support for power connections.
- *State Electricity Regulatory Commissions* are accountable for concessional electricity rates, open access, and other regulatory incentives or support for battery swapping services.

The vehicle segments targeted for battery swapping are e-2Wheelers and e-3Wheelers, which are highly concentrated in metropolitan areas. Consequently, the rollout of battery switching stations will be staggered as follows:

- **Phase 1 (Years 1-2):** All metropolitan cities with a population of more than 4 million (according to the 2011 Census) would receive priority for implementing battery swapping networks during the first phase.
- **Phase 2 (Years 2-3):** All significant cities, including state capitals, UT headquarters, and cities with populations exceeding five lakhs (according to the 2011 Census), would be included in the second phase will give the importance of the 2W and 3W vehicle segments in rising cities.

When providing additional policy assistance and subsidy allocations for battery swapping networks, central and state government bodies participating in the development of battery swapping networks may take this ranking into account.

Battery Swapping: The US and China Experience and Learnings for India.

There are high barriers to entry in the EV business - hefty cost of batteries, investments in charging and power infrastructure, and expansion of the electric grid. These are some of the major reasons why the adoption of EVs has not taken off in the US. According to Pradeep K. Chintagunta, professor of marketing at The University of Chicago Booth School of Business[4] "*Since EVs use emerging battery technologies, they face several significant technical, economic, and social barriers to adoption, limiting EV penetration in the US*". To overcome this problem, the US government has complemented tax credits and subsidies to consumers with a *$100* billion investments towards infrastructure upgrades, as well as an investment of *$174* billion to build half a million charging stations by 2030 (Anjani Trivedi, 2021).

China on the other hand has seen a wider adoption of EVs in all the segments - 2, 3, and 4 wheelers. At the core of it strategy has been policies to reduce the barriers to EV adoption. The outcome has been that China has nearly *1.68 million* charging points, half of which have been publicly set up. It has one charging point for every five EVs, the corresponding number for US is one charging point for every *20* EVs. In addition to subsidies, China has focused its efforts on improving electricity grid networks - it has plans to invest nearly *$900 billion* over the next five years and improving the availability and accessibility of Battery Swapping Stations.

Battery Swapping as a service is steadily gaining prominence. A battery is the most significant cost item in an EV comprising nearly *45 - 50%* of the total cost of the vehicle. Battery swapping lowers the cost of purchase of a vehicle as the consumer can now purchase the EV without the battery and then pays for the energy used. Several players have entered the battery swapping business. For example, Nio, China's leading EV manufacturer recently opened its *700th* battery swapping station and has plans to increase the number of such station to approximately *4000* by 2030 (Lewin Day, 2022). Nio has a subscription model where a user pays approximately *USD 150* for a 70 KWh battery. The world's largest battery maker, Contemporary Amparex Technology Co. Ltd (CATL) is setting up facilities to set up swap stations in 10 cities across China. To this extent it has come up with a new battery swap solution called EVOGO comprising of battery blocks, fast battery swaps and an App (Murray Slovick, Feb, 2022). However, battery swapping has not be a success everywhere. In the US, Tesla entered the battery swapping business in

2013, but then eventually exited the business in two years. An Israeli startup BetterPlace had set up a network of battery swapping stations in 2012, but eventually filed for bankruptcy in 2013 (Lewin Day, 2022).

The above findings hold important lessons for India. Incentivizing consumers alone is addressing only a piece of the EV ecosystem. Unless the charging accessibility and power grid infrastructure are put in place, EV manufacturers as well as well as consumers will have little interest in participating in the EV market. In this regard the Government of India in its Union Budget for 2022-23 listed improving the Electric Vehicle (EV) ecosystem to spur the demand for green vehicles, and as part of this exercise has come up with a Battery Swapping Policy. The Finance Minister of India, Nirmala Sitharaman said in her budget speech (Sitharaman, 2022) on Feb 01, 2022.

"Considering the constraint of space in urban areas for setting up charging stations at scale, a battery swapping policy will be brought out and interoperability standards will be formulated. The private sector will be encouraged to develop sustainable and innovative business models for **'Battery or Energy as a Service'**.*"*

One of the major objectives of the policy is to encourage the adoptions of Battery swapping as a service that can encourage more homegrown startups to emerge thereby lowering the cost of adoption to the consumer. Finally, Original Equipment Manufacturers (OEMs) still prefer integrating the battery with the vehicle instead of having standardized designs where the battery is decoupled from the sale of the vehicle. This practice will have an impact on the EV ecosystem, and therefore will influence the Circular Economy.

ACKNOWLEDGMENT

We thank Chandrashekhar Bhide, Co-Founder, Lithion Power agree for sharing his experiences regarding the startup venture and for his valuable inputs related to the EVs ecosystem.

REFERENCES

Ahmadi, L., Yip, A., Fowler, M., Young, S. B., & Fraser, R. A. (2014). Environmental feasibility of re-use of electric vehicle batteries. *Sustainable Energy Technologies and Assessments*, *6*, 64–74. doi:10.1016/j.seta.2014.01.006

Ahuja, J., Dawson, L., & Lee, R. (2020). A circular economy for electric vehicle batteries: driving the change. *Journal of Property, Planning and Environmental Law*.

Anjani Trivedi. (2021). *What America can learn from China on electrifying its traffic.* Mint. Retrieved from https://www.livemint.com/opinion/columns/what-america-can-learn-from-china-on-electrifying-its-traffic-11618759172498.html

CarDekho. (2021), *Top EV-friendly States In India That Offer The Best Incentives To Electric Car Buyers.* https://www.cardekho.com/india-car-news/top-evfriendly-states-in-india-that-offer-the-best-incentives-to-electric-car-buyers-27251.htm

D'Amato, D., Droste, N., Allen, B., Kettunen, M., Lähtinen, K., Korhonen, J., & Toppinen, A. (2017). Green, circular, bio economy: A comparative analysis of sustainability avenues. *Journal of Cleaner Production*, *168*, 716–734. doi:10.1016/j.jclepro.2017.09.053

Day. (2022). *China loves battery swapping EVs, but will they ever make it here?* Hackaday. Retrieved from https://hackaday.com/2022/02/09/china-loves-battery-swapping-evs-but-will-they-ever-make-it-here/

De Giovanni, P. (2021). Digital Supply Chain Through IoT, Design Quality, and Circular Economy. In *Dynamic Quality Models and Games in Digital Supply Chains* (pp. 57–89). Springer. doi:10.1007/978-3-030-66537-1_4

De Giovanni, P. (2022). Leveraging the Circular Economy with Closed-loop Supply Chain and Reverse Omnichannel using Blockchain Technology and Incentives. *International Journal of Operations & Production Management*, *42*(7), 959–994. doi:10.1108/IJOPM-07-2021-0445

De Giovanni, P., & Folgiero, P. (2022). Strategies for the Circular Economy: Circular Districts and Networks. *Routledge-Giappichelli.*

DelhiS. (n.d.). https://ev.delhi.gov.in/charging-station

Dobrotă, D., & Dobrotă, G. (2018). An innovative method in the regeneration of waste rubber and the sustainable development. *Journal of Cleaner Production*, *172*, 3591–3599. doi:10.1016/j.jclepro.2017.03.022

Economic Times. (2021). *India's Top Innovative MSMEs*. https://economictimes.indiatimes.com/small-biz/top-innovative-msme/msme_companies/categoryid-01.cms

Hill, N., Clarke, D., Blair, L., & Menadue, H. (2019). Circular economy perspectives for the management of batteries used in electric vehicles. Final Project Report by Ricardo Energy & Environment for the JRC. *Publications Office of the European Union, 537140*. Advance online publication. 10

Hirst, D., Butcher, L., & Hinson, S. (2018). *Electric vehicles and infrastructure*. House of Commons Library. . doi:10.1007/978-3-540-33037-0_13

Kaza, S., Yao, L., Bhada-Tata, P., & Van Woerden, F. (2018). *What a waste 2.0: a global snapshot of solid waste management to 2050*. World Bank Publications. doi:10.1596/978-1-4648-1329-0

Kirchherr, J., Reike, D., & Hekkert, M. (2017). Conceptualizing the circular economy: An analysis of 114 definitions. *Resources, Conservation and Recycling*, *127*, 221–232. doi:10.1016/j.resconrec.2017.09.005

Landrum, N. E., & Ohsowski, B. (2018). Identifying worldviews on corporate sustainability: A content analysis of corporate sustainability reports. *Business Strategy and the Environment*, *27*(1), 128–151. doi:10.1002/bse.1989

Maranesi, C., & De Giovanni, P. (2020). Modern circular economy: Corporate strategy, supply chain, and industrial symbiosis. *Sustainability*, *12*(22), 9383. doi:10.3390u12229383

National Institutional for Transforming India (NITI). (2022). https://www.niti.gov.in/draft-battery-swapping-policy

Sitharaman, N. (2022). *Government of India Budget 2022-2023*. https://www.indiabudget.gov.in/doc/budget_speech.pdf

Slovick. (2022). *Electronic Design, Battery-Swapping Stations Pick Up the Pace in China*. https://www.electronicdesign.com/markets/automotive/article/21216583/electronic-design-batteryswapping-stations-pick-up-the-pace-in-china

World Economic Forum. (n.d.). *From linear to circular—Accelerating a proven concept.* https://reports.weforum.org/toward-the-circular-economy-accelerating-the-scale-up-across-global-supply-chains/from-linear-to-circular-accelerating-a-proven-concept/

Chapter 3
Metaverse:
A New Platform for Circular Smart Cities

Behzad Maleki Vishkaei
https://orcid.org/0000-0003-1297-4710
Luiss University, Italy

EXECUTIVE SUMMARY

The way we interact with the physical world around us is being changed rapidly by advances in technology. These new technologies are very helpful for developing smart cities with the goal of improving the quality of life considering sustainability. Technology is also at the core of the circular city vision, and we need to use the new technologies and act in a smarter way in smart cities to support the achievement of economic, environmental, and social targets that are important for sustainable development and circular economy (CE) implementation. The metaverse (which is the combination of the prefix "meta" with the word "universe") is a new concept that is known as transcending hypothetical synthetic environment linked to the physical world. In this chapter, the proposed concept will be introduced in more detail, and then the authors discuss how smart cities can use the metaverse platforms for improving themselves in five main aspects including governance, environment, mobility, economy, and quality of life dimensions.

INTRODUCTION

The metaverse is one of the most up-to-date technological developments in the rapidly transforming world in this regard. It is valuable to observe and

DOI: 10.4018/978-1-6684-5001-7.ch003

follow the equivalent of the discussions in this direction in the literature (Damar, 2021). Stephenson (1992) for the first time mentioned the metaverse in a piece of speculative fiction named Snow Crash defining it as a massive virtual environment parallel to the physical world, in which users interact through digital avatars. Lee et al. (2021) provided a comprehensive framework that examines the latest metaverse development under the dimensions of state-of-the-art technologies and metaverse ecosystems, and illustrates the possibility of the digital 'big bang'.

Shi et al. (2004) presented the initial design of such an end-to-end transport service for metaverse applications, along with the results of a simulation study evaluating its effectiveness. Bourlakis et al. (2009) studied the evolution of retailing, i.e. from traditional to electronic to metaverse retailing, and shed light on the ways metaverses influence that evolution considering the key challenges and opportunities faced by traditional retailers, e-retailers, and metaverse retailers. Vernaza et al. (2012) diussed the research that has been developed at the University of Panama for the use of the metaverse (virtual worlds) based on free software as virtual learning environments and their applications in e-Learning retaled to Electronics. Nevelsteen (2018) obtained a definition for a "virtual world" via sample technologies using grounded theory and compared it with related work and used it to classify advanced technologies such as a pseudo-persistent video game, a MANet, virtual and mixed reality, and metaverse. Thomason (2021) studied how the metaverse may be used in the future to change, enhance, and possibly transform health care considering collaborative working, education; clinical care, wellness, and monetization. Duan et al. (2021) provided a three-layer metaverse architecture including infrastructure, interaction, and ecosystem.

The metaverse is an immersive platform and an embodied internet where you can experience everything that you can imagine in a virtual world. In other words, users live within a digital universe via using various technologies such as telepresence, virtual reality, and augmented reality. It takes a few years before the key features of the metaverse become mainstream but already some of the high-tech companies in the world are working on platforms that can be a part of the metaverse in the near future.

Meta company (formerly Facebook) envisions a virtual world where digital avatars connect through work, travel, or entertainment using VR headsets. Acquiring Oculus (a provider of virtual reality equipment) in 2014 by this company is considered as a step toward its goal regarding metaverse in the future. To combine the real world with augmented reality and virtual reality, Microsoft company is developing mixed and extended reality (XR) applications with its Microsoft Mesh platform. Epic Games company is working on a photorealistic digital human's platform (MetaHuman Creator) that allows you to customize your digital doppelganger in future open world games. Some sample figures are depicted from the proposed companies in Figures 1,2 and 3 respectively.

Figure 1. Virtual World (Source: Meta Company)

Figure 2. Azur Mixed Reality (Source: Microsoft company)

Figure 3. Metahuman Creator (Source: Epic Games company)

Based on research by Lee et al. (2021), to build the metaverse we need to move from the Physical World to Digital Twins and Finally create the metaverse.

Digital Twins is a large-scale and high-fidelity digital model and entities (instead of a single object such as nut and bolt) that is duplicated in a virtual

Figure 4. Physical World, Digital Twins, and Metaverse (Source: Meta company)

environment and is connected to its Physical Twins by their data. It includes a physical counterpart with various properties such as object motions, and temperature. The metaverse could exist as a self-sustaining and persistent virtual world that coexists and interoperates with the physical world with a high level of independence. The metaverse is able to support unlimited numbers of users, or avatars in a number of virtual worlds to experience heterogeneous activities in real-time. Figure 4 simply shows Physical World, Digital Twins, and Metaverse.

Based on research by Lee et al. (2021), the metaverse ecosystem includes six pillars including Avatar, Content creation, Virtual economy, Social acceptability, Security & privacy, and Trust & accountability. Moreover, the metaverse technology enablers consist of eight pillars counting Network, Edge/Cloud, Artificial Intelligence, Computer Vision, Blockchain, Robotics/IoT, User Interactivity, and Extended Reality.

As many of the mentioned technologies are used in smart cities, it seems the metaverse as a new concept and platform cannot be ignored in the future of smart cities. The metaverse will be a new platform for smart cities and implementing of a circular economy vision which leads to increase the economic productivity through reduced congestion, eliminated waste, and reduced costs. It also will be helpful for improving the quality of life and the economy of the city by helping to reduce pollution, enhancing social interaction, and providing new opportunities for prospering different businesses.

The smart city idea was born from the application of hi-tech solutions to urban problems, especially from the use of ICT in connecting people, political institutions, and businesses (Dameri, 2014). In other words, a smart city is meant to be actively engaged in improving the quality of life of its citizens and in pursuing sustainable growth, thanks to the wide use of ICT (Fontana, 2014).

Anthopoulos (2015) developed a literature review to discover and classify the particular schools of thought, universities, and research centers as well as companies that deal with the smart city domain. Meijer and Bolivar (2016) studied smart city governance as a complex process of institutional change and acknowledged the political nature of appealing visions of sociotechnical governance. Angelidou (2017) discussed whether and in what way characteristics of smart cities are present in the plans of 15 cities: Amsterdam, Barcelona, London, PlanIT Valley, Stockholm, Cyberjaya, Singapore, King Abdullah Economic City, Masdar, Skolkovo, Songdo, Chicago, New York, Rio de Janeiro, and Konza. Camero and Alba (2019) provided research to explore the computer science and information technology literature about Smart City. Caragliu and Del (2019) analyzed the urban innovation impact of smart city policies. Laufset al. (2020) did a systematic review to explore the recent literature concerned with new 'smart city' security technologies aiming to investigate to what extent these new interventions correspond with traditional functions of security interventions.

Eight different classes of smart city conceptualization models have been discovered, which structure the unified conceptualization model and concern smart city facilities (i.e., energy, water, IoT, etc.), services (i.e., health, education, etc.), governance, planning, and management, architecture, data and people (Anthopoulos et al., 2019). Noori et al. (2021) developed an Input-Output model to help policymakers and analysts to further their conceptual understanding of smart cities, envisage design choices they will face during planning and implementation, and help them to understand the impact of these choices. They also illustrated the model by introducing the case of "Smart Dubai."

The main features of a smart city include a high degree of information technology integration and a comprehensive application of information resources. The essential components of urban development for a smart city should include smart technology, smart industry, smart services, smart management, and smart life (Kim et al., 2017). Yeh (2017) surveyed citizens in Taiwanese cities and revealed that citizens are willing to accept and use ICT-based smart city services if the services are designed with innovative concepts that secure their privacy and offer a high quality of services. Santi et al. (2018) studied the smart city vision, providing information on the main requirements and high-lighting the benefits of integrating different Internet of Things (IoT) ecosystems within Cloud under this new Cloud of Things (CoT) vision. The results of the study of Yigitcanlar et al. (2021) revealed that: (a) innovation, sustainability, and governance are the most popular smart

Figure 5. Dimensions and factors for assessment of smart cities based on Ortega-Fernández et al. (2020)

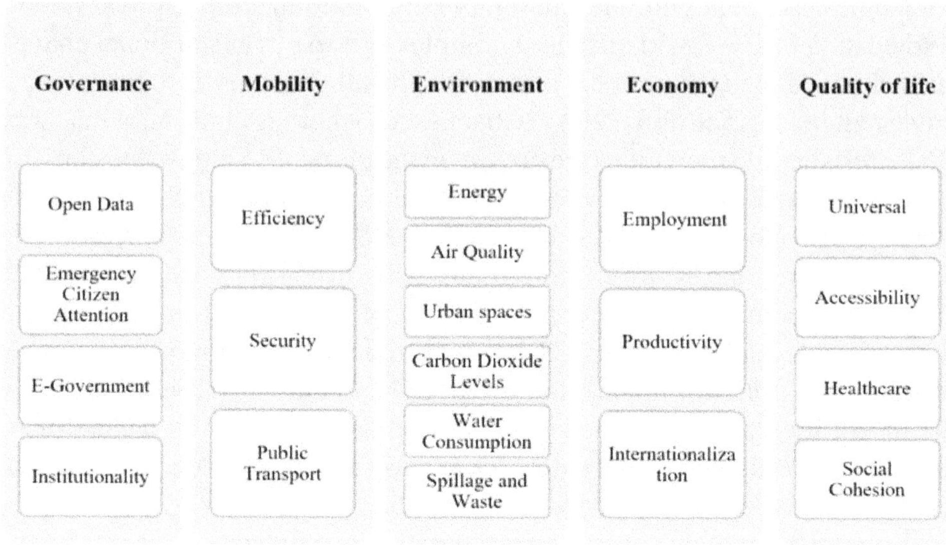

city concepts; (b) internet-of-things, artificial intelligence, and autonomous vehicle technology are the most popular technologies; (c) a balanced view exists on the importance of both smart city concepts and technologies.

Smart systems make daily activities more easily, efficient and represent real support for sustainable city development. To improve the quality and performance of cities is recommended to involve all interested parties to implement efficiently smart systems (Batagan, 2011). Smart cities can be evaluated in different aspects considering various dimensions, factors, and indicators. Ortega-Fernández et al. (2020) considered six main dimensions and 20 factors for analyzing the performance of smart cities. Figure 5 shows the proposed dimensions and their related factors for smart city assessment.

The governance dimension mostly focuses on adopting transparent decision making, providing online services, and encouraging citizens to more participate by enabling helpful platforms. The mobility dimension is related to creating an integrated mobility system that is safe and accessible to everyone considering the environmental impact. The environment dimension concentrates on renewable energy and energy saving and also reducing environmental pollution such as greenhouse gases emission. The economic dimension is related to supporting entrepreneurship and innovation to provide a flexible labor market for everyone especially women and young people.

The last dimension (quality of life) tries to improve public health and welfare besides enhancing the education system. It also concentrates on civil issues such as improving tourist policies, promotion of social cohesion, and civic engagement.

In the next sections, it will be discussed how the metaverse can have an important role in helping smart cities to improve themselves in five main dimensions including governance, environment, mobility, economy, and quality of life.

METAVERSE AND GOVERNANCE

Governments are responsible for providing public services and the metaverse can bring the public services online at a high-quality level. Using online communities, many of the services can be provided to the public sector in a more efficient way. Many of the bureaucratic processes can be managed or eliminated via virtual communication without any need to stay in a long queue and more accurate data is gathered and stored for further analysis with the goal of improving public services. Citizens may have access to needed data considering privacy rules and cooperate in making new decisions regarding their city in a virtual community. Many of the services can be simulated in the virtual world to find out about their advantages and disadvantages before implementing them in the real world. The cost of constructing new buildings for different communities can be decreased by providing virtual environments.

Based on the announcement by the Seoul Metropolitan Government (SMG), it seems this city is on the path to becoming the first major city to enter the metaverse and it is developing a comprehensive long-term plan for a metaverse. The SMG is planning to shift consultations and civil services to a virtual public service center staffed by the avatars of public officials by the name “Metaverse 120 Center”. VR headsets will be used by the citizens to meet city officials and they also will be able to attend mass events.

Despite there are patchworks of privacy laws such as the EU’s General Data Protection Regulation (GDPR), and the California Consumer Privacy Act (CCPA) to protect citizens, the metaverse may increase the privacy concerns, and governments should devise and use new standards to assure their citizens they will be protected regarding their privacy.

Figure 6. Seoul Mayor Oh Se-hoon attends an event as an avatar in a metaverse, in a demonstration of the city's plans (Source: Seoul City Government)

METAVERSE AND MOBILITY

According to the data that is released on the Statista website, U.S. travelers took 464 million domestic business trips in 2019 (https://www.statista.com/statistics/207103/forecasted-number-of-domestic-trips-in-the-us/). As estimated, domestic business travels in the U.S. dropped to 185 million in 2020 as a result of the travel restrictions due to the coronavirus (COVID-19) pandemic. It is forecast that the number of business trips will recover gradually during the following years, reaching 457 million by 2024. Although the shared mobility system such as public bicycle sharing is known as a new workable transportation system for reducing pollution and traffic in cities during rush hours and helping people to gain healthier lives, they could not cut down unnecessary daily travels (Vishkaei et al., 2020).

In the future because of metaverse platforms, business travels for the purposes of making eye contact and shaking hands becomes wasteful insanity. The metaverse will eliminate the time lost and energy spent in transit and the employees or students can gather at their virtual workplace or virtual schools respectively to avoid unnecessary commuting. Moreover, many of the trips that are related to leisure time will also be ignored by people as they can attend different events like concerts via entering the metaverse to save their time and cost. Regarding physical or real transportation, the metaverse can be helpful for simulating roads, vehicles, and transportation systems to analyze customers' experiences before developing new projects.

Recently, the metaverse or metauniverse have been called by the car manufacturers and a new concept of 'metamobility' is born by Hyundai Motor. As in the future robotics will act as a medium to connect the virtual and real worlds, automobiles and other mobility vehicles including Urban Air Mobility (UAM) will be smart devices to access virtual spaces. For example, an automobile that connects to virtual spaces can allow users to enjoy various in-car VR experiences such as having a trip to Mars with the help of the robots that are placed on the proposed planet. Depending on the user's needs, a car can be transformed into an entertainment space, a meeting room for work, or even a 3D video game platform (HYUNDAI press, 2022).

Figure 7. The experience of traveling to Mars while being transferred in a metamobility vehicle (Source: Hyundai press)

METAVERSE AND ENVIRONMENT

These days, smart cities are trying to use sustainable mobility services to reduce fossil fuel consumption and consequently greenhouse gasses (Vishkaei et al., 2021). But these actions are not sufficient to reach the main goals of sustainable development. According to the previous section, the metaverse can be useful for reducing the number of daily trips which leads to a significant reduction in air pollution especially the pollution related to the greenhouse gases that are emitted into the air by cars, airplanes, and other mobility vehicles. Therefore, emissions from commuting will be cut. Employees will be available in the virtual offices so the number of real buildings that are currently occupied by various private and governmental offices will decrease. This will also result in fewer energy consumptions including electricity and gas that are being consumed by these buildings.

Based on the report of Statista website (https://www.statista.com/statistics/276480/world-carbon-dioxide-emissions-by-sector/) the energy and heat production sector are responsible for much of the world's carbon dioxide (CO2) emissions. In 2018, this sector emitted approximately 14 billion metric tons of CO2. This was almost the combined total produced by both the transport and manufacturing sectors. The metaverse and the virtual world may be a solution for decreasing a part of carbon dioxide emissions.

Although implementing the metaverse can help us to reduce pollution, e-waste will remain one of the main concerns. The metaverse works by massive data holding which is now accessible through cloud computing. It also requires new technologies and facilities such as sensors, headsets, and superfast networks. Currently, the world is facing the e-waste problem which is produced in an amount of more than 50 million tons per year. Therefore, it is vital to devise new standards that prevent producing e-waste because of using the new technologies and facilities that are required for metaverse platforms.

METAVERSE AND ECONOMY

The metaverse will be a marketplace where creators can sell and share 3D digital items and services besides selling their physical goods. On the one hand, consumers need to develop skills and capabilities, change habits, and rethink product usability. On the other hand, companies need to invest in the latest technologies to create smart products that complement their current value proposition with new features, insights, and applications (Cesaretto et al., 2021). Good environmental performance pushes up sales, thus becoming an interesting marketing lever (Liu & De Giovanni, 2019).

The metaverse is the virtual world that people and companies are able to run their businesses in different fields and of course, there would be other types of businesses that are directly related to the metaverse existence such as working as a Metaverse Architecture and Designer, Ecosystem Developer, and Metaverse Storyteller. To run a business, shopping, and even open an account on metaverse platforms, crypto wallets will be needed, which will hold all of your digital currency. The type of cryptocurrency depends on which platform you are interested to use.

Based on a report that was released by Bloomberg Intelligence in 2021, the global metaverse revenue opportunity could approach $800 billion in 2024 vs. about $500 billion in 2020 (https://www.bloomberg.com/professional/

Figure 8. Metaverse market growth outlook (Source: Bloomberg Intelligence, Newzoo, IDC, PWC, Statista, and Two Circles data)

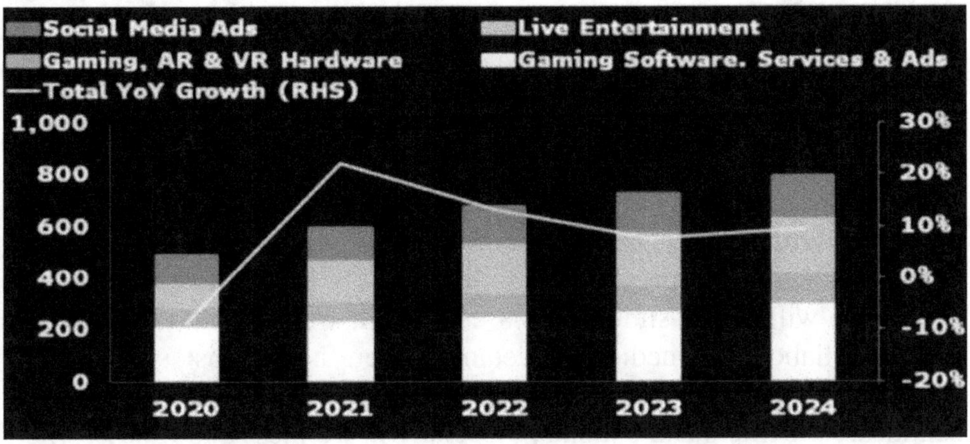

blog/metaverse-may-be-800-billion-market-next-tech-platform/). The primary market for online game makers and gaming hardware may exceed $400 billion in 2024 while opportunities in live entertainment and social media make up the remainder.

Moreover, Specialists from different countries and regions can gather in their high-quality virtual workplaces and all the companies will be able to hire international experts from anywhere. All of these may help smart cities to create new job opportunities and enhance their productivity in different fields and industries.

METAVERSE AND QUALITY OF LIFE

Obviously, as the metaverse can help us to improve the Governance, Mobility, Environment, Education, Health, and Economy systems, it can be effective for enhancing the quality of our lives.

The metaverse can encourage and enable collaboration despite the country or region you are living in. Social presence is generally defined as the awareness of being present with others in a mediated environment combined with a certain degree of attention to the other's intentional, cognitive, or affective states (Biocca & Harms, 2002; Green & Taber, 1980). Moreover, people in virtual worlds also experience copresence because they feel they are in a world together (Biocca et al., 2002).

Figure 9. Social presence in Metaverse (Source: Meta Company)

In the virtual world, people can collaborate and have an avatar-based interaction via social presence and the ability to manipulate avatars. The Metaverse will allow simultaneous education, training, and planning as well as collaborative medical procedures. There is immense scope for the metaverse to be used in clinical care. Gamification is a new way of connecting healthcare providers and patients, especially in wellness and fitness, where AR can deliver smarter workouts with guidance from virtual instructors. The metaverse will be used for surgical simulations, diagnostic imaging, patient care management, rehabilitation, and health management (Van der Land et al., 2011).

By the metaverse, patients can receive more realistic consultations using avatars, and they can also be beneficiaries of more effective treatment and diagnosis systems through data interconnectivity and digital twins.

Figure 10. Physical and mental activities in Metaverse (Source: Meta Company)

The virtual world can be expanded to the education systems using avatars, virtual classrooms, virtual labs, and conferences. The use of metaverse expands to college events and social activities in virtual space. Seoul National University Medical School utilized metaverse in anatomy practice in the first semester of 2021. This applies VR and AR to anatomy, etc. with MedicalIP, a company specializing in artificial intelligence medical image (di Bella et al., 2014).

Figure 11. Hana Bank employees take a finance class organized by their firm in the metaverse (Source: Hana Bank)

METAVERSE AND THE CIRVULAR ECONOMY

The concept of circularity has been described as a tangible driver of innovations and value creation for the 21st century global economy, dealing with the complex issues of waste, economic growth, and sustainability (Greyson, 2007). The European Circular Economy Action Plan identifies CE as an essential element of a broader transformation towards climate neutrality and long-term competitiveness. Indeed, CE makes it possible to combine environmental sustainability with economic sustainability (De Giovanni, 2020). Typical futuristic smart cities focus on applying circular economy principles, particularly in the energy sector, in terms of resource utilization and waste mitigation aspects (Musti, 2020). Moreover, firms invest in circular economy programs for performing the maximum possible return rate, which is possible only through an efficient digitalization process (Maranesi & De Giovanni, 2020). Smart cities try to enhance the quality of life considering sustainable development. The new technologies and platforms such as the metaverse are the key tools for being smarter and achieving sustainable goals.

Circular Economy might be placed at the intersection of the ecological, economic, and social dimensions of sustainability (Suárez-Eiroa et al., 2019). It

Figure 12. Relationship between circular economy and sustainable development

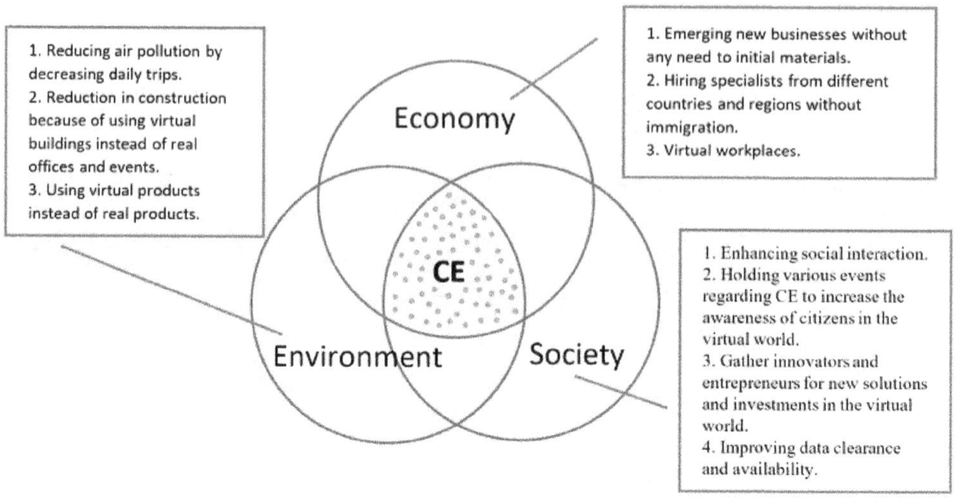

is extremely complex for a city to define which combinations of CE initiatives will help achieve its sustainability goals. When approaching circularity, cities might look for assessments that deal with CE and the strategies that they seek to implement (Petit-Boix & Leipold, 2018). Circular Economy provides the goals and principles, and smart solutions and the new technologies and platforms in the smart cities help to deliver them. Figure 12 indicates how the metaverse is connected to different dimensions of sustainable development and consequently the Circular Economy.

The circular economy is a regenerative system in which resource input and waste, emission, and energy leakage are minimized by slowing, closing, and narrowing material and energy loops. Based on Figure 12, metaverse can minimize energy consumption via eliminating unnecessary construction, using virtual workplaces to reduce daily trips, devising new business models without consuming too many materials, and providing virtual products. Moreover, increasing people's awareness, enhancing data clearance and availability, and encouraging entrepreneurs to invest in CE projects are other necessary points that can be managed more efficiently using the metaverse platform.

CONCLUSION

To summarize the main points mentioned in this chapter, it would be wise for local governments to take into account the metaverse as a new platform that can be helpful for smart cities. They can realize more potential cost savings or gains through the implementation of the metaverse and this would be an effective way to improve the services provided to the people, particularly in vital sectors such as government, environment, health, education, transportation, and communications.

Moreover, as there is a strong relationship between the dimensions of sustainable development and Circular Economy, and one of the main goals of smart cities is enhancing sustainability, the metaverse can also be known as a platform that can be useful for implementing CE strategies in a smart city.

In the near future, the metaverse will be used to change, enhance, and possibly transform our daily lifestyle and smart cities seem to be pioneers in considering this platform as a helpful solution to act smarter. The metaverse is introduced as another enormous entity parallel to our physical reality but we know there are still many challenges to be overcome before the metaverse becomes integrated into the physical world which can be achieved using the new technologies that are emerging these days or will be developed in the near future. Therefore, we expect to see more research outcomes in this field regarding new coming technologies.

ACKNOWLEDGMENT

This contribution is a part of a project that has received funding from the Lazio region ESF operational programme 2014-2020, Axis III- Education and Training, Investment Priority 10.ii) - Specific Objective 10.5, Pivotal Action 21 – Call for proposals “Contributi per la permanenza nel mondo accademico delle eccellenze” – Project Title: “Implementazione di progetti di digitalizzazione responsabile nelle supply chain” - CUP (Unified Project Code): F86J20002710009.

REFERENCES

Angelidou, M. (2017). The role of smart city characteristics in the plans of fifteen cities. *Journal of Urban Technology*, *24*(4), 3–28. doi:10.1080/10630732.2017.1348880

Anthopoulos, L., Janssen, M., & Weerakkody, V. (2019). A Unified Smart City Model (USCM) for smart city conceptualization and benchmarking. *Smart cities and smart spaces: Concepts, methodologies, tools, and applications*, 247-264.

Anthopoulos, L. G. (2015). Understanding the smart city domain: A literature review. *Transforming city governments for successful smart cities*, 9-21.

Batagan, L. (2011). Indicators for economic and social development of future smart city. *Journal of Applied Quantitative Methods*, *6*(3), 27–34.

Biocca, F., & Harms, C. (2002). What is social presence? In F. Gouveia & F. Biocca (Eds.), *Presence 2002 Proceedings*. University of Fernando Pessoa Press.

Bourlakis, M., Papagiannidis, S., & Li, F. (2009). Retail spatial evolution: Paving the way from traditional to metaverse retailing. *Electronic Commerce Research*, *9*(1), 135–148. doi:10.100710660-009-9030-8

Camero, A., & Alba, E. (2019). Smart City and information technology: A review. *Cities, 93*, 84-94.

Caragliu, A., & Del Bo, C. F. (2019). Smart innovative cities: The impact of Smart City policies on urban innovation. *Technological Forecasting and Social Change*, *142*, 373–383. doi:10.1016/j.techfore.2018.07.022

Cesaretto, R., Buratto, A., & De Giovanni, P. (2021). Mitigating the feature fatigue effect for smart products through digital servitization. *Computers & Industrial Engineering*, *156*, 107218. doi:10.1016/j.cie.2021.107218

Damar, M. (2021). Metaverse Shape of Your Life for Future: A bibliometric snapshot. *Journal of Metaverse*, *1*(1), 1–8.

Dameri, R. P. (2014). Comparing smart and digital city: initiatives and strategies in Amsterdam and Genoa. Are they digital and/or smart? In *Smart city* (pp. 45–88). Springer.

De Giovanni, P. (2020). *Dynamic quality models and games in digital supply chains*. Springer Nature.

di Bella, E., Odone, F., Corsi, M., Sillitti, A., & Breu, R. (2014). Smart Security: Integrated systems for security policies in urban environments. In *Smart city* (pp. 193–219). Springer.

Du, R., Santi, P., Xiao, M., Vasilakos, A. V., & Fischione, C. (2018). The sensable city: A survey on the deployment and management for smart city monitoring. *IEEE Communications Surveys and Tutorials*, *21*(2), 1533–1560. doi:10.1109/COMST.2018.2881008

Duan, H., Li, J., Fan, S., Lin, Z., Wu, X., & Cai, W. (2021, October). Metaverse for social good: A university campus prototype. In *Proceedings of the 29th ACM International Conference on Multimedia* (pp. 153-161). 10.1145/3474085.3479238

Fontana, F. (2014). The smart city and the creation of local public value. In *Smart City* (pp. 117–137). Springer.

Green, S. G., & Taber, T. D. (1980). The effects of three social decision schemes on decision group process. *Organizational Behavior and Human Performance*, *25*(1), 97–106. doi:10.1016/0030-5073(80)90027-6

Greyson, J. (2007). An economic instrument for zero waste, economic growth and sustainability. *Journal of Cleaner Production*, *15*(13-14), 1382–1390. doi:10.1016/j.jclepro.2006.07.019

Kim, T. H., Ramos, C., & Mohammed, S. (2017). Smart city and IoT. *Future Generation Computer Systems*, *76*, 159–162. doi:10.1016/j.future.2017.03.034

Laufs, J., Borrion, H., & Bradford, B. (2020). Security and the smart city: A systematic review. *Sustainable Cities and Society*, *55*, 102023. doi:10.1016/j.scs.2020.102023

Lee, L. H., Braud, T., Zhou, P., Wang, L., Xu, D., Lin, Z., ... Hui, P. (2021). All one needs to know about metaverse: A complete survey on technological singularity, virtual ecosystem, and research agenda. *Journal of Latex Class Files*, *14*(8).

Liu, B., & De Giovanni, P. (2019). Green process innovation through Industry 4.0 technologies and supply chain coordination. *Annals of Operations Research*, 1–36. doi:10.100710479-019-03498-3

Maranesi, C., & De Giovanni, P. (2020). Modern circular economy: Corporate strategy, supply chain, and industrial symbiosis. *Sustainability*, *12*(22), 9383. doi:10.3390u12229383

Meijer, A., & Bolívar, M. P. R. (2016). Governing the smart city: A review of the literature on smart urban governance. *International Review of Administrative Sciences*, *82*(2), 392–408. doi:10.1177/0020852314564308

Musti, K. S. (2020). Circular economy in energizing smart cities. In *Handbook of Research on Entrepreneurship Development and Opportunities in Circular Economy* (pp. 251–269). IGI Global. doi:10.4018/978-1-7998-5116-5.ch013

Nevelsteen, K. J. (2018). Virtual world, defined from a technological perspective and applied to video games, mixed reality, and the Metaverse. *Computer Animation and Virtual Worlds*, *29*(1), e1752. doi:10.1002/cav.1752

Noori, N., de Jong, M., Janssen, M., Schraven, D., & Hoppe, T. (2021). Input-output modeling for smart city development. *Journal of Urban Technology*, *28*(1-2), 71–92. doi:10.1080/10630732.2020.1794728

Ortega-Fernández, A., Martín-Rojas, R., & García-Morales, V. J. (2020). Artificial intelligence in the urban environment: Smart cities as models for developing innovation and sustainability. *Sustainability*, *12*(19), 7860. doi:10.3390u12197860

Petit-Boix, A., & Leipold, S. (2018). Circular economy in cities: Reviewing how environmental research aligns with local practices. *Journal of Cleaner Production*, *195*, 1270–1281. doi:10.1016/j.jclepro.2018.05.281

Shi, S., Wang, L., Calvert, K. L., & Griffioen, J. N. (2004, April). A multi-path routing service for immersive environments. In *IEEE International Symposium on Cluster Computing and the Grid* (pp. 699-706). IEEE.

Stephenson, N. (1992). *Snow crash: A novel.* Bantam Books.

Suárez-Eiroa, B., Fernández, E., Méndez-Martínez, G., & Soto-Oñate, D. (2019). Operational principles of circular economy for sustainable development: Linking theory and practice. *Journal of Cleaner Production*, *214*, 952–961. doi:10.1016/j.jclepro.2018.12.271

Thomason, J. (2021). MetaHealth-How will the Metaverse Change Health Care? *Journal of Metaverse*, *1*(1), 13–16.

Van der Land, S., Schouten, A., & Feldberg, F. (2011). Modeling the metaverse: A theoretical model of effective team collaboration in 3D virtual environments. *Journal of Virtual Worlds Research*, *4*(3). Advance online publication. doi:10.4101/jvwr.v4i3.6126

Vernaza, A., Armuelles, V. I., & Ruiz, I. (2012, June). Towards to an open and interoperable virtual learning enviroment using Metaverse at University of Panama. In 2012 Technologies Applied to Electronics Teaching (TAEE) (pp. 320-325). IEEE. doi:10.1109/TAEE.2012.6235458

Vishkaei, B. M., Fathi, M., Khakifirooz, M., & De Giovanni, P. (2021). Bi-objective optimization for customers' satisfaction improvement in a Public Bicycle Sharing System. *Computers & Industrial Engineering*, *161*, 107587. doi:10.1016/j.cie.2021.107587

Vishkaei, B. M., Mahdavi, I., Mahdavi-Amiri, N., & Khorram, E. (2020). Balancing public bicycle sharing system using inventory critical levels in queuing network. *Computers & Industrial Engineering*, *141*, 106277. doi:10.1016/j.cie.2020.106277

Yeh, H. (2017). The effects of successful ICT-based smart city services: From citizens' perspectives. *Government Information Quarterly*, *34*(3), 556–565. doi:10.1016/j.giq.2017.05.001

Yigitcanlar, T., Kankanamge, N., & Vella, K. (2021). How are smart city concepts and technologies perceived and utilized? A systematic geo-Twitter analysis of smart cities in Australia. *Journal of Urban Technology*, *28*(1-2), 135–154. doi:10.1080/10630732.2020.1753483

Chapter 4
Ethical Phone for a Fair, Circular, and Sustainable Future:
Fairphone Business Case and Possible Application in a Smart City Context

Sara Romagnoli
BSC Consulting, Italy

Sofia Brunelli
A2A, Italy

Camilla Bruscino
Deloitte, Italy

Alessio Ciotti
LabGov.City, Italy

Sofia Di Giò Pietro
PwC, Italy

Tommaso Dumontel
LabGov.City, Italy

Chiara Fusari
Agricolus S.r.l., Italy

Francesca Nacca
Gorillas, Italy

EXECUTIVE SUMMARY

Fairphone is a Dutch company that produces ethical mobile phones relying on a circular business model. The company was born after a campaign condemning the "bloody minerals" sourced by the mobile phone industry and the huge amount of e-waste this industry produces. For what attains the reduction of e-waste, Fairphone is implementing both slowing and closing resource loop strategies to reduce the tendency of rapid smartphone replacements. Fairphone is also engaging in complex supply chain management operations to assure that all its partners provide high quality social standards. Given the complexity of the business model, Fairphone has created a unique network of stakeholders,

DOI: 10.4018/978-1-6684-5001-7.ch004

with customers playing a pivotal role, each providing fundamental help. Given the innovative and holistic approach Fairphone is implementing, the company could also generate positive effects in the growing field of smart cities. The chapter is dedicated to the analysis of Fairphone's business case and the positive impacts the company could generate within smart cities.

INTRODUCTION

For the past 150 years (Reuter et al., 2018), many of the products we use in our everyday life have been manufactured according to the principles of a linear economy: new or existing raw materials are extracted, processed into a ready-to-consume product and finally thrown away after use, thus becoming waste. This process forces the economy to repeat the same pattern over and over again: extraction, production, distribution, consumption and disposal. This causes enormous environmental damage, but in recent years there has been a growing awareness of the need to move towards a circular economy: an economy designed to regenerate itself, so that waste becomes a resource for others. According to the Ellen MacArthur Foundation, the circular economy approach aims to maximize the usefulness of products, their components and materials throughout their life cycle, recovering inputs once products reach the end of their useful life.

Circular economy has been implemented in many sectors, including the tech industry. The latter generates an increasing amount of waste year over year, furthermore, only 18% of such waste (e-waste) has been collected and recycled in 2019 (Barros & Dimla, 2021). The production system traditionally used by the tech industry is planned obsolescence. This is a strategy adopted by companies in order to make a product that will become obsolete in a set time frame (De Giovanni, 2022). Planned obsolescence determines the creation of a huge amount of electronic waste having a negative impact on the environment: in fact, its disposal requires a specific process due to the presence of materials such as lead, copper and mercury that, if disposed of in the wrong way, can seriously harm the environment. This phenomenon has generated awareness and the need to approach a circular economy.

The tech industry is approaching the transition from linear to circular economy both rethinking product design and component use (Jalali et al., 2020). The main challenge is ensuring a long-life product with higher performance components thus reducing e-waste.

In this paper, we discuss and analyze the benefits of the circular economy in the phone industry, focusing on the Fairphone's case, a company founded in the Netherlands in 2013 led by Bas van Abel. We have chosen to present it as a business case for its integral and innovative approach to the circular economy. Fairphone managed to become a company capable of confronting bigger and globalized multinational companies that still focus more on the profit rather than ethical principles.

Fairphone was founded with the aim of creating a fairer and more sustainable phone industry by motivating people and businesses to act more responsibly, as we will discuss in the following paragraphs, in particular in section 4. But how is this done?

The products used by the company are chosen to ensure increased sustainability and adequate working conditions. This particular aspect will be discussed in subsequent sections (1-2-3-5), analyzing the various actors involved within the Fairphone organization, how the company tailored its strategy on ethical and circular models, and which transformations are being carried out.

Fairphone is currently trying to invert the trend of replacing a smartphone about every 18 months, as this generates a huge environmental impact. Fairphone's design is modular, upgradable and durable, allowing users who purchase this phone to be able to repair it more quickly and easily through tutorials provided by the company itself.

The paper will then focus on another very important aspect, covered in section 6: the possible implications on smart cities. Smart cities, analyzed in section 6, include a set of strategies that involve human, intellectual and social capital, with the goal of making daily activities easier and more efficient and assure a sustainable city development (Vishkaei et al., 2021). As the digital transition is progressing, it has brought many changes including an increased use of smartphones, which many systems have integrated to connect citizens with the city (GPS navigation, the ability to see real-time traffic, access to public services, etc..).

CIRCULAR ECONOMY NETWORK

Fairphone is about discovering new production systems, tackling global issues and stimulating debate about what is fair. Fairphone is trying to generate positive impacts in two main fields: firstly, on the entire value chain from the extraction of minerals, through the design, production and life cycle of the telephone; secondly, on the relevant market, by supplying ethical products.

When analyzing Fairphone's circularity network, we see that the company decided to address both the key aspects of circularity (Bocken et al, 2016):

- Slowing resource loops: extend the useful life of the produced phones
- Closing resource loops: the use of recycled materials within its value chain.

Slowing Resource Loops

Regarding the first point, Fairphone is planning to extend the useful life of the smartphone as much as possible, the aim is an average period of three years.

One of the pillars of the Fairphone project is reparability: the phone is made with easily removable modules, allowing consumers to repair and possibly replace them autonomously. For this reason, Fairphone is working on a service network able to quickly send spare components wherever required in Europe.

High standard materials are essential in creating a long-lasting high-quality product, and Fairphone is willing to link quality and transparency within its supply chain including all the actors involved, such as mines, foundries, refineries and different levels of producers (De Giovanni, 2021).

The company is trying to map the suppliers according to their social and environmental standards, selecting only those suppliers with the highest ones. Fairphone has identified one hundred and three suppliers, fully involved in the mission of the company to promote a greater understanding of the issues at stake. According to the company's policy, the relationship with suppliers aims to identify opportunities to create a positive change. Therefore, relationships are based on collaborations, evaluations, and improvement programs, such as the creation of the Fairtrade gold alliance and the Fair cobalt alliance. The attention is expanded also to manufacturers, and the company outsources the manufacturing of the components only to industries respecting the TCO standards for adequate working conditions (Fairphone,2020). The company engages also with governments, experienced NGOs in order to promote development projects for local communities in the underdeveloped countries where mining and manufacturing plants are located (Falcone, 2018).

Figure 1. Fairphone network

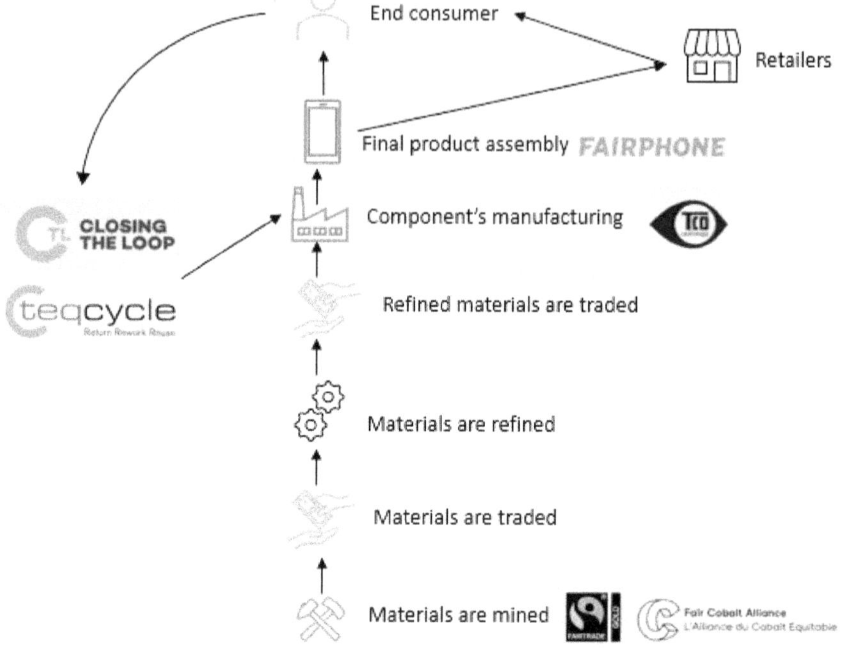

Closing Resource Loops

Fairphone aims to increase the percentage of recycled and recyclable material in all the phone's components. In doing so the company cooperates with two main partners: Teqcycle and Closing the loop.

Teqcycle is a German company based in Munich that deals with private companies and public administrations for the collection, recycling and proper disposal of hardware components in exhausted phones. Fairphone and Teqcycle are involved in the Recycling Program which foresees the collection of no longer usable phones, their disassembly and the separate reuse or re-selling of the raw materials and components. The program also involves consumers who do not own a Fairphone, which are engaged in returning their old devices.

The other important partner of Fairphone is Closing the Loop, a Dutch company which collects exhausted devices to extract plastic. The two companies together have collected 3.1 tons of discarded phones and recycled them properly (Fairphone, 2016).

ANALYSIS OF THE VALUE, DRIVERS, STAKEHOLDERS, AND BARRIERS

"We care about people and the planet", this is the Mission of Fairphone, and, as hard as it may seem, they really managed to do it within the value creation of the company.

Value: Fairphone identifies social and environmental weaknesses in the electronics industry, and it is committed to continuous performance improvement and global efforts to create a more livable world. All Key Performance Indicators contribute, directly or indirectly, to the United Nations Sustainable Development Goals - the agenda collectively agreed by humanity to build a fairer and more sustainable future - and call on industry peers to ensure that their goals align with this important work. Emotional, physical and technical durability all play a role in the longevity of smartphones. Fairphone, thanks to its sustainable values, wants to enable the user to keep their phone for at least 5 years thus postponing the product obsolescence. The use of the circular economy inside the business process contributes significantly to the reduction of CO2 emission by reducing the manufacturing of brand-new phones. Overall, Fairphone sets four main objectives: long-lasting design, reuse and recycling, fair materials and good working conditions (Fairphone, 2021). To analyze more in detail the specifications of the value, we can subdivide into Value Proposition, Value Creation and Value Capture as follow:

Value Proposition: Fairphone is committed to offering an alternative smartphone with do-it-yourself repairs, customization options, software upgrades and it allows for long-term replacement of parts. The main objective is to ensure and sustain the longevity of the phone and thus leverage the lower environmental impact of the device itself. (Circular X, 2020)

Value Creation: From 2013 to 2021, Fairphone has produced no less than 4 new smartphone models and it is also selling eight repair modules for the phone. For replacement work the user has an online repair kit, spare parts directory and the Fairphone online community available at all times. (Fairphone, 2021)

Value Capture: Fairphone 4 retails at 579€. Spare parts and accessories are also available to buy from Fairphone, with the costs of repair considerably cheaper than repairs for non-modular phones directly from the official website. Fairphone also takes back old phones, offering a cash back.

Table 1. Drivers

Internal	External
Desire to improve the industry for the benefit of society	Scale of the problem to impact
Improve or develop a technology	Environmental and social crisis
Network embedment	Market failure
Managerial perspective	Market opportunities
Organizational identity	Customer demand
Organizational capabilities	Community support
Intangible assets	Social expectations
Being a role model	NGOs and stakeholder pressure
	Knowledge exchange

Drivers: Fairphone experienced a market for fair and sustainable products, in which many people will pay a premium for an ethical product, and that there are no unsurmountable technical or logistical barriers to the pursuit of fair materials and practices. Materials research and strategic partnerships incentivize customers to give back old phones and optimize Fairphone take-back and repair logistics.

When analyzing the drivers, is it possible to divide them between drivers for social entrepreneurship for sustainable innovations on one side and internal performance drivers on the other side.

Fairphone began with a social mission driven by a desire to improve the electronics industry by making it more socially and environmentally responsible. Over the years, the number of drivers has increased. The following are the social entrepreneurship drivers for Fairphone's sustainable innovation. They are divide into 8 internal and 9 external drivers. (Bruzi, 2019)

While analyzing the internal performance drivers, Fairphone takes into account the following aspects which will be analyzed more in detail in section 4:

1. Willingness of people to pay for ethical products
2. % Phones Sold
3. % Phones in use vs. sold
4. % Phones recovered vs. sold
5. Average % of 8 focus materials sustainably sourced
6. People benefiting from Fairphone's social interventions
7. Industry influence score
8. % of used material that can be recovered

9. Post-consumer recycled materials from electronics (Fairphone, 2021)

Finally, another important driver for Fairphone is the "right to repair" which has been recently identified within the main legislative priorities of the European Green Deal. This right gives consumers the ability to repair their own devices thus fostering the ecofriendly impact of the company.

Stakeholders

A multi-stakeholder approach is key to Fairphone's theory of change and fair sourcing strategy. The company is committed to working with its suppliers and other stakeholders (such as industry peers, nonprofits and governments) to create coalitions that develop fair and transparent supply chains that benefit workers and communities. In addition, Fairphone is very active in Europe in actively engaging customers in returning their old devices.

Moreover, to achieve a sustainable and responsible mining supply chain, Fairphone focuses its partnership with artisanal and small-scale mining (ASM).

Fairphone was the first electronics company to source gold from Fairtrade certified artisanal mines. These mines have improved working conditions and received a premium for the gold they produce. Fairphone has integrated this gold into the Printed Circuit Board (PCB) supply chain for the previous Fairphone 3 model. (Fairphone, 2021)

Other key partners are the ones mentioned above and connected to the collection of old smartphones such as Teqcycle and Closing the Loop.

But in the Fairphone Network the main stakeholders are the customers. Fairphone relies a lot on customer experience and customer engagement by promoting the Fairphone Ambassador Program and a Forum community. By implementing a more equitable business model and creating strategic partnerships with key players in the industry and with sustainable values, Fairphone helps motivate the electronics industry to also engage in caring for people and the planet.

Barriers

Fairphone has had more difficult challenges than expected in convincing some of its component suppliers to approach their suppliers for alternative sources. Sometimes, the company has had to go four, five and six levels deep into a supplier's supply chain. And often there is no straight business case for a sub-supplier to collaborate in this journey, as “we are usually the first clients to ask” (Fairphone, 2021). This highlights the importance of scaling sustainable source suppliers and normalizing the expectation of, and demand for, sustainable material supply chains. The company also faces major product related challenges in design for recycling and those are reported as follow (Reuter et al., 2018):

Limits

- There is a design/functionality limitation regarding the number of different modules that can be easily disassembled and have in the one product.
- When carrying out recycling, its design must take into account that combinations of metals, compounds, fillers, plastics and functional materials are required to minimize the loss of elements and waste. These complex combinations with many different materials existing in telephones make this a very challenging process.
- For the realization of a successful design, the process must rely on a lot of accessible and consistent data. However, this accessibility and disclosure is not always guaranteed at industry or supplier level.
- Recycling and repairability go hand in hand.

HOW THE CIRCULAR ECONOMY IMPACTS THE CORPORATE STRATEGY

According to Maranesi and De Giovanni (2020), the corporate strategy can be defined as the area within the business activities where the fundamental decisions are taken (De Giovanni, 2016). More in particular, the decisions that commit the company over a medium to long time horizon, under conditions of uncertainty, to acquire additional shares of the economic value in the specific sector.

Applying this definition to the Fairphone company, it is evident that circularity embeds the whole corporate strategy, pervading every single choice made within the company. In this regard, it is worth mentioning that most companies are addressing sustainability in various ways, which often end up as an “add-on” to traditional business practices (De Giovanni and Vinzi, 2014). Fairphone, instead, decided to enter in the market as a social enterprise and created its own business in order to address the social and environmental needs identified, from the very beginning (Wernik et al., 2014).

The same history of the company brings further evidence of this peculiar approach. In 2010 the future company leader Von Abel started a campaign to raise awareness on social issues related to mineral extraction in African countries. The campaign was focused on the use of "bloody mineral" in the technology industry, and the huge amount of e-waste generated. Starting from the success of the initiative, Von Abel started to collaborate with the charity entity Waap to create the first "fair" smartphone. In 2013 Fairphone decided to take a commercial approach, creating an independent social enterprise (Fairphone, 2021).

It is evident how in the Fairphone case circularity was not a second step decision taken to respond to external pressures (such as political will or marketing campaign) but was instead one of the pillars of the production of the "fair" phone.

This is mirrored in a series of choices of corporate strategy taken by the company:

1. Fairphone decided to remain independent from investors in the first years of activity, opting to finance the launch of the first fairphone with a crowdfunding campaign. More than 25,000 people believed in the project and bought the smartphone for €325 in a pre-sell, making its production possible. When the company obtained enough financial stability decided to accept investments only from socially aware shareholders (impact investments) and, given the relatively small finance available and the willingness to reduce e-waste, decided to adopt a non-continuous production flow: the production only starts when a certain number of phones has been pre-ordered.
2. Fairphone decided to adopt a complete transparency policy: the company's finances are completely disclosed, as well as the contracts and relationship with the manufacturer and the suppliers.
3. Fairphone decided to design a targeted communication and advertising strategy putting greater efforts in communicating with targeted consumers and having an "activism-oriented" approach. Communication activities not only aim to present the product, but also to raise awareness of the issues of technological waste and the exploitation of coltan, gold and other materials in mining. In this way Fairphone manages to awaken the consciences of consumers and advocate for an active participation in the collection of old smartphones and other electronic devices.

Figure 2. Abell Model

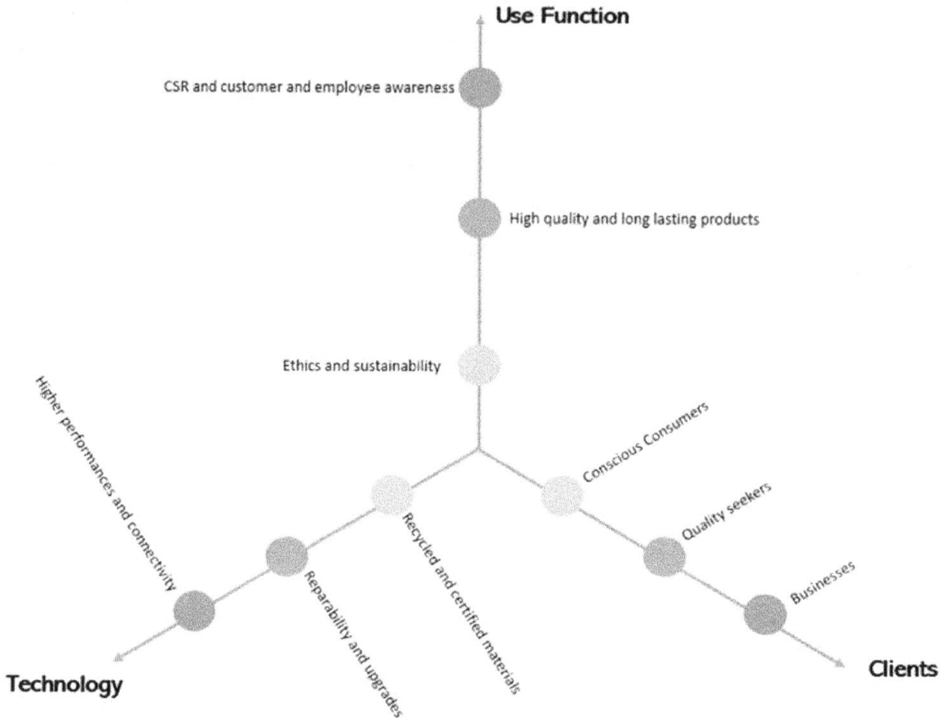

4. Fairphone decided to differentiate the market by launching a specific campaign for businesses. The possibility to sell the product directly in bigger slots pre-ordered by companies that choose the Fairphone as a business phone fits perfectly with Fairphone's production chain.

To better illustrate the corporate strategy, it is possible to adopt the Abell Model. Regarding Fairhone's core business, the selling of smartphones, we can identify three main Business Strategy Areas:

- Yellow area: Fairphone started to address the so-called "conscious consumers", identifiable as shoppers who are increasingly choosing to make conscious buying decisions. The main characteristic for this segment is the ethical and sustainable appeal of the Fairphone which is provided through the use of certified and recycled materials.
- Green area: Fairphone is trying to attract customers particularly interested in the quality of the products, the so-called Quality Seekers.

These clients look for high quality and long-lasting products, which Fairphone is producing thanks to the important upgrades achieved in the last two models. The main drivers to attract these clients is the reparability and the possibility to upgrade both the phone and the software.

- Blue area: Businesses are the latest segment that Fairphone is willing to attract, preparing targeted advertising campaigns and providing discounts for companies choosing fairphones as business phones. These companies are attracted by the possibility to implement an indirect CSR campaign (also through the promotion done by Fairphone itself) and nudge their own employees. The technology driver in this case is mainly the higher performances of the last models (e.g., Fairphone 4 has the 5G connection, improved cameras and storage capacity).

The links between secular economy system and performance Corporate social responsibility refers to the efforts that a company makes to take responsibility for its own actions by considering the impact that those may have on the environment and society. But the key question is: is this business choice also economically feasible?

To answer this question, we mainly used the data reported into different reports published by Fairphone that show the information in a clear and simple way, making comparisons with previous years and considering also the objectives that the company had set itself in previous years.

Since 2013, Fairphone has been building a business centered on fairness. Today, Fairphone proves that this is not only possible, but also is economically sustainable. Fairphone increased its revenues by 87% gaining more and more importance within the electronic industry, as it is shown in the company's 2020 Impact Report (Fairphone, 2021).

The Fairphone Impact Report is distributed yearly to measure the organization's advancement towards the objective of creating the fairest and environmentally friendly product it can and rousing the business to act more responsibly. The report shows the most recent goals that impact on the organization's KPIs:

- **Number of telephones sold**: The volume of phones sold bounced by 76%. Fairphone sold nearly 95,000 products including accessories, modules, and extra parts - contrasted with the 53,000 devices in 2019 (Ibid).

- **Percentage of telephones recuperated versus sold in Europe**: In 2020, Fairphone set the objective of making up for at least 14% of Fairphones sold by gathering old gadgets through its takeback program and it surpassed this objective: the organization made up for 18% of Fairphones sold. For 2021, the organization focuses on 45% and will incorporate the assortment of e-waste outside the EU (Fairphone's impact 2020: A challenge to the industry).
- **Average percentage of 8 focus materials fairly sourced**: Around 56% of Fairphone's eight focus materials have been fairly obtained, rather than the 25% in 2018 and 32% in 2019. The 8 focus materials were picked in light of the fact that they are the materials that offer the electronic industry the best potential to make a positive effect. From this year on, Fairphone will build its aspirations to coordinate 70% of 14 Fair materials by 2023 (Fairphone's impact 2020: A challenge to the industry).
- **Number of people benefiting from Fairphone's social interventions**: 2020 saw 10,717 individuals profiting from Fairphone's social intercessions, up from 5,296 out of 2018 and 7,839 of every 2019. This KPI reflects the number of people who have profited from Fairphone's social, ecological or potentially financial interventions in mines and industrial facilities (Fairphone's impact 2020: A challenge to the industry).
- **Industry impact**: Fairphone keeps on driving a more extensive effect inside the electronic business, as its impact develops every year. Despite the fact that impact is hard to measure, Fairphone right now measures its industry impact dependent on a framework that relegates focuses to organizations that follow Fairphones' initiatives dependent on their size, influencer and market size. In 2020, Fairphone accomplished 31 impact focuses in view of its joint efforts with organizations including Glencore, Tesla, and/e/OS, a critical figure contrasted with 10 of every 2018 and 13 out of 2019 (Fairphone's impact 2020: A challenge to the industry).

To highlight the important results that the company achieved in 2020 Eva Gouwens, CEO at Fairphone states: "*Each year, our sales figures have steadily grown and in 2020 we turned a profit for the first time since 2014. We are pleased to have finished the year in the black, missing our pre-Covid sales goal, but doing much better than expected. This is a significant milestone for us and we want to continue on this upward financial trajectory. By establishing*

Figure 3. Source: "Fairphone's impact 2020: A challenge to the industry"

	Fairphone Impact KPIs 2020	Goal	Actual
1	# Phones Sold	110,373	94,985
2	% Phones in use vs. sold	68%	60.5%
3	% Phones recovered vs. sold	14%	18.1%
4	Average % of 8 focus materials sustainably sourced	70%	56%
5	# People benefiting from Fairphone's social interventions	9.000	10.717
6	Industry influence score #	18	31

a viable market for ethical consumer electronics, we want to inspire the entire industry to act more responsibly. To be an inspiring example, we need to show that we're financially sustainable and commercially successful, while making more ethical choices in our business and supply chain" (Fairphone, 2021)

Still, when analyzing the performances an important issue needs to be addressed. Fairphone's business model is able to provide very little margins for the company itself. We were able to find precise data on cost breakdown only for the Fairphone 2 but we can assume that this data is similar for the later models. From the report "Cost breakdown of the Fairphone 2" (Fairphone, 2021), the company declared that its margin obtained from the sale of a single phone is 9€ on a sale price of 525€. This is equal to 1.7%, which is very low when confronted with other smartphones: iPhone 7 granted Apple with the 42% of margins and Samsung Galaxy S7 with the 32% (Dedrik et al., 2019). The low value of this margin can be explained by different aspects:

Figure 4. Source: Fairphone: "Cost breakdown of the Fairphone 2"

- Taxes and Reseller Margin. Almost 23% of the phone's total price is absorbed by the taxes (VAT and WEEE) and resellers so the real amount that Fairphone can use to make the final output is 407€.
- Product. 65% of the phone's price is used to produce the product itself.
- Investment. 6% of the total price is invested in R&D to create high-performance phones and to promote social innovation
- Operations. 5% of the total price is used to cover operating costs such as administration and customer service that is handled in house.

The remaining margin is then reserved to face either unexpected costs or further investments in social impact programs and product development.

CIRCULAR ECONOMY DEVELOPMENTS AND BUSINESS TRANSFORMATION

As we have previously seen the mobile phone industry is still characterized by a linear model of production based on planned obsolescence. Planned obsolescence (PO) can be defined as "the conscious decision taken by a company to produce a product that will become obsolete in a predefined time frame" (Barros & Dimla, 2021). There are three types of planned obsolescence: of function, of quality and of desirability (Ibid).

Obsolescence of function occurs when a new product outperforms an existing one. Companies in this case release a product in the market while planning the next one in the meantime. This is the most common strategy and consists in the update of components such as the camera and the music player.

Obsolescence of quality, or technology obsolescence, occurs when a product is not performing anymore due to a faulty or broken component. The lifecycle of this component has been intentionally reduced by design. Furthermore, components usually have different life cycles making the repairing process even more difficult. It is important to stress that for the mobile phone industry, technology obsolescence also refers to softwares. A new software can make a previous one obsolete, but it is not only about softwares *per se*. An update in the software is also capable of making the hardware obsolete. A good example of a company implementing a technology obsolescence strategy is Apple in it is iPhone (Ibid). Apple not only develops the operating system but also the App Store where customers can download third party apps. Therefore, Apple can implement a technological obsolescence strategy: by designing products so that they are difficult to be repaired by third parties, by updating the software making hardware obsolete and lastly by intervening on the firmware (Ibid). Firmware allows the configuration and interoperability of hardware components. In this context Apple uses the firmware to prevent unauthorized repairs and limit product usability (Ibid). This strategy has been recently applied to batteries and is now being applied to other components such as the camera. For example, if the battery is replaced, the firmware prompts error messages and disables the battery capacity reader together with other functions, such as the Face ID reader (Ibid). Similarly, if the camera is replaced, the firmware limits camera functionality.

The last type of planned obsolescence is obsolescence of desirability. This strategy consists in changing superficial aspects of the product with the aim of making the previous product undesirable. In this case the core function of the product is still sound. For this reason, this obsolescence strategy is also called psychological or aesthetic obsolescence. An example of this strategy is the one implemented by Nokia between 2004 and 2006 (Ibid). In 2004 Nokia released the first fashion collection reinterpreting the Art Deco movement through the lens of high tech. In 2005 Nokia released the "L'Amour Collection" which was defined as trend-inspired. Furthermore, these collections have been released during a fashion show underlying even more the emphasis on aesthetics.

As we have seen, this model of production generates adverse consequences for society and the environment. Fairphone is challenging this model of production with an innovative business model capable of tackling many of the issues created by the linear model based on planned obsolescence. We will now take a look at Fairephone's business model highlighting the strategies that make its business circular.

There are two fundamental strategies for the cycling of resources (Bocken et al, 2016):

1) Slowing resource loops: through the design of products that are meant to last for a longer time span and by offering product life extension services. This strategy allows customers to use the product more hence the flow of resources slows down.
2) Closing resource loops: through recycling hence closing the loop between post-use and production

For what attains the slowing the resource loop strategy, Fairphone focuses more specifically on design for product-life extension (Ibid). "This strategy is concerned with the extension of the use period of goods through the introduction of service loops to extend product life, including reuse of the product itself, maintenance, repair, and technical upgrading, and a combination of these" (Ibid). This strategy can be divided into four sub-strategies (Ibid):

- Design for maintenance and repair: Fairphones' products present a modular design making them easier to repair (Nußholz, 2018). Furthermore, Fairphone makes single components and repair guides easily available to customers (Ibid).
- Design for upgradability and adaptability: Thanks to the modular design it is possible to upgrade single components without changing the others.

- Design for standardization and compatibility: this strategy aims at creating products with parts or interfaces that fit other products as well (Ibid). As we have just seen, Fairphone allows this.
- Design for dis- and reassembly: Fairphon's modular design clearly allows this

For what attains the closing the resource loop strategy, Fairphone follows two sub-strategies, Design for Technological cycle and Design for Disassembly and Reassembly. The former refers to a design of the product that allows a safe and continuous recycle of the materials used (Bocken et al, 2016). The latter instead refers to a design of a product that facilitates the separation of components hence the recycling process (Ibid).

As we have seen Fairphone pursues both a slowing and closing resource loop strategy. Of course, this aspect influences its business model. The latter has both the characteristics of an "encourage sufficiency business model" (Ibid), typical of companies pursuing a slowing the resource loop strategy, and of an "extending resource value business model" (Ibid), typical of companies pursuing a closing the resource loop strategy. The former refers to solutions aiming at the reduction of end-user consumption "through principles such as durability, upgradability, service, warranties and reparability and a non-consumerist approach to marketing and sales" (Ibid). The latter, instead, refers to solutions aiming at "exploiting the residual value of resources: collection and sourcing of otherwise "wasted" materials or resources to turn these into new forms of value" (Ibid). Given the peculiarity of Fairphone's business model, the following paragraphs are dedicated to the description of it.

To guarantee a prolonged use of the product the company, as we have previously seen, focuses on a modular design and on making available to all customers repair guides. Fairphone's value proposition includes high level performance standards, reliability, and low life cycle costs (Nußholz, 2018). Fairphone's customers are both individuals and public or private organizations. What is interesting to underline in relation to customers is that Fairphone aims to create a community among them (Ibid). It is possible to access certain services thanks to an account on the website, participate in Q & As and organize events with other Fairphone users. Crucial activities for Fairphone, as we have previously seen, are management of value chain activities, sustainability standards implementation and operations such as sales and shipping (Ibid). Key resources and capabilities, instead, are just-sourced minerals, the skills required to produce a competitive and modular phone and a transparent and sustainable supply chain (Ibid). Fairphone has key partners in order to carry out its business, these include, components manufacturers, software developers but also industry networks and NGOs which support Fairphone in the implementation of sustainability standard programs (Fairphone, 2018). The Dutch Company then sells its products through its website and through third-party retailers. Costs are represented by the production and operations while revenues are generated by the sale of products.

Fairphone's business model has been conceived also considering the collection and reintegration of end-of-use mobile phones, this is not limited to Fairphone's products, but it is extended to all types of mobile phones (Duhaylongsod and De Giovanni, 2018). Phones are shipped by the customers themselves hence this activity does not represent a cost. Although it is important, for Fairphone, to foster the community-based approach with customers since it represents a powerful incentive to ship an end-of-use phone (Nußholz, 2018). Key activities in this case are motivating people to ship their end-of-use phones and informing them about practicalities such as whether the phones can be either reused or recycled (Fairphone, 2018). Key resources and capabilities here are represented by an adequate return system. In this phase it is very important to underline the partnership Fairphone has with Teqcycle. Costs result from the collection process and no revenues are generated (Nußholz, 2018).

Teqcycle is responsible for the disassembly part and when performing this activity on Fairphones the process is more efficient thanks to the modular design. At this stage recovered parts are either sold to third parties or, if possible, reused for the production of Fairphones (Ibid). Costs result from the disassembly process and revenues are generated thanks to the sale of recovered parts (Ibid). This phase also leads to the refurbishment of end-of-use mobile phones. But these are sold by Teqcycle hence this activity is not considered within Fairphone's business model (Sacco and De Giovanni, 2019).

Fairphone also offers the recovery of materials such as gold and glass (Fairphone, 2018). Once again, the key resource here is the modular design that allows very high material recovery rates (Reuter et al, 2018). Also, in this case the partnership with Teqcycle plays a crucial role. The secondary raw materials are either sold or reused in the production of new Fairphones. Costs result from the recycling activities and revenues are generated by the sale of secondary raw materials.

One last thing that is worth mentioning in relation to Fairphone's business model is that it can be considered to some extent open (Kortman & Piller, 2016). This is because consumers participate in its business model at different

Figure 5. Source: Purece, Cristian. (2019). IRIS -Integrated and Replicable Solutions for Co-Creation in Sustainable Cities

stages. Consumers participate in design contests and local production enabled by 3D printers (Ibid). But what is more interesting is that consumers are able to provide financial resources thanks to crowdfunding campaigns hence assuming the responsibilities of classical shareholders (Ibid).

POSSIBLE IMPLICATIONS IN TERMS OF SMART CITIES

In this paragraph we will show the link between smart cities and Fairphone. Even if it is not glaring, Fairphone's circular business model can foster e-waste reduction and improving waste management in smart and circular cities.

A Smart City System comprises of six key building blocks: (i) smart people, (ii) smart city economy, (iii) smart mobility, (iv) smart environment, (v) smart living, and (vi) smart governance (Vinod Kumar & Dahiya, 2017). Moreover, the notions of smart cities and smart economy address the problems related to cities and try to enhance their planned and sustainable development, efficient management, and effective and participatory governance. Typically, the economy of a smart city is increasingly driven by technically inspired innovation, creativity and entrepreneurship thanks to smart people. So, the concept of smart city is built on a combination of ideas on how ICTs might contribute to improvements in the functioning of cities, improving their competitiveness, enhancing their efficiency, and finding new ways to tackle problems of poverty, social deprivation, and poor environmental and energy management. Indeed, it is not surprising that the notion of smart city directly relates to the concept of sustainable urban development.

At the same time, the arrival of smartphones in the last decades has boosted the Information Age and the world of ICTs. A standard smartphone can support both basic telephone features (Internet, e-mail and audio-visual communication) and all sorts of digital applications, music and movie player, camera, voice dictation, GPS navigation, among other things (De Giovanni, 2019).

Given the above premises, an important concept related to smart city is the one of real-time city. "Smart urbanism envisages a thoroughly digital city in which city services, infrastructures and populations are managed in real-time using ICTs" (Kitchin, 2017). In this context smartphones play a key role. Real-time mobile and location-based social networking and journey planner smartphone apps seek to provide flexibility and improve efficiency.

An interconnected and intelligent city system enables the capture and integration of live real-world-data through the use of sensors, meters, personal devices, the web, appliances, camera, smartphones and another similar data-acquisition system including social networks as networks of human sensors interconnected.

As part of the transition towards sustainability, the development of future urban systems includes creating not just smart cities but also circular cities with a focus on the well-being of the entire ecosystem. The principles of circular economies (recycling, reusing, repairing, refurbishing, remanufacturing to reduce waste and the environmental impact) are unified with the idea that urban systems should be smart and sustainable (Cialani et al, 2021). It is in this context that Fairphone's added value stands out. Given the importance of both circular economy and smartphones in smart cities, Fairphone could be a means to improve the quality of life and make daily activities more easy, efficient and it represents a real support for sustainable city development from a circular perspective (De Giovanni, 2022).

Moreover, since Fairphone is based on a circular business model, both waste reduction and recycling are intrinsic values of the company, which is very in line with the core concept of smart and circular city. One of the key principles of circular cities is turning waste into a resource. This can be achieved if citizens are engaged and highly committed to separate waste, but also if they favor recycling processes as much as they can and choosing a smartphone like Fairphone could be a great step forward. Furthermore, Fairphone tries to reduce e-waste by recycling old phones that people send back, which is coherent with the primary goal of circular cities.

Finally, although the concept of big data is very important in a smart city, so is the concept of privacy. Smart cities use sensors and other data collection tools to amass vast banks of information that are then analyzed and processed. In light of this, it is important to underline that Fairphone has a new partnership with /e/ Foundation for an open-source operating system that prioritizes user privacy since also their clients' privacy is one of their key concerns (Fairphone, 2021).

CONCLUSIONS

The aim of this paper was to provide an in-depth analysis of Fairphone's business case. We began with analyzing the main activities carried out by Fairphone in order to provide a general overview of the company. We then focused on Fairphone's peculiar network of stakeholders underlying all the operations the company carries out to assure a transparent and sustainable supply chain. Section 2, instead, was dedicated to the analysis of how Fairphone generates and captures value, the drivers and barriers, both internal and external, faced by Fairphone and the main stakeholders Fairphone engages with. Section 3 provides a thorough analysis of how Fairphone's strong beliefs craft its corporate strategy in a holistic manner. We then concentrated on providing an overview of Fairphone's performance mainly drawing upon the reports disclosed by the company. As we have seen Fairphone is rapidly growing especially in terms of sales, but it is important to stress that due to the high costs encountered, their margins are relatively low. In section 5 we focused on Fairphone's business model in order to demonstrate Fairphone's integrated approach to circular economy, pursuing both slowing and closing resource loop strategies. The last section was dedicated to analyzing the positive effects Fairphone can generate in the context of smart cities, especially focusing on the disposal of e-waste. We can conclude that Fairphone is playing a pivotal role in paving the way towards a circular tech industry with an innovative and comprehensive approach that focuses not only on environmental sustainability, but also on the social one. Fairphone is universally considered a virtuous example of circularity and environmental and social commitment. Our analysis shows a new entrepreneurial path which could be a real game changer not only in the tech industry, but also in the future development of our society, as we have shown in section 6 referring to the smart cities. We recognize the exceptionality of the Fairphone case given its peculiar story and company's value proposition, and we wonder whether such characteristics represent a condition sine qua non an enterprise could be considered really "fair", and whether traditionally structured companies are willing and able to embark on a similar path. It will be interesting to see whether Fairphone will manage to push even further to the circularity of the tech industry by producing new goods in a fair and sustainable manner.

REFERENCES

Barros, M., & Dimla, E. (2021). From planned obsolescence to the circular economy in the smartphone industry: An evolution of strategies embodied in product features. *Proceedings of the Design Society, 1*, 1607–1616. doi:10.1017/pds.2021.422

Bocken, N. M., De Pauw, I., Bakker, C., & Van Der Grinten, B. (2016). Product design and business model strategies for a circular economy. *Journal of Industrial and Production Engineering, 33*(5), 308-320.

Bruzi, A. (2021). Social Entrepreneurship For Sustainable Innovations: A Case Study Of Fairphone. *Zeszyty Programu TOP, 15*, 11.

Circular, X. (2020). *Case study: Fairphone - Modular, long-life smartphones.* https://www.circularx.eu/en/cases/36/fairphone-modular-long-life-smartphones

De Giovanni, P. (2016). Coordination in a distribution channel with decisions on the nature of incentives and share-dependency on pricing. *The Journal of the Operational Research Society, 67*(8), 1034–1049. doi:10.1057/jors.2015.118

De Giovanni, P. (2019). When feature-based production capabilities challenge operations: Drivers, moderators, and performance. *International Journal of Operations & Production Management, 40*(2), 221–242. doi:10.1108/IJOPM-04-2019-0309

De Giovanni, P. (2020). *Dynamic quality models and games in digital supply chains*. Springer Nature.

De Giovanni, P. (2022). Leveraging the circular economy with a closed-loop supply chain and a reverse omnichannel using blockchain technology and incentives. *International Journal of Operations & Production Management, 42*(7), 959–994. doi:10.1108/IJOPM-07-2021-0445

De Giovanni, P., & Vinzi, V. E. (2014). The benefits of a monitoring strategy for firms subject to the Emissions Trading System. *Transportation Research Part D, Transport and Environment, 33*, 220–233. doi:10.1016/j.trd.2014.06.008

De Giovanni, P., & Zaccour, G. (2022). A selective survey of game-theoretic models of closed-loop supply chains. *Annals of Operations Research, 314*(1), 77–116. doi:10.100710479-021-04483-5

Dedrick, J., & Kraemer, K. L. (2017). *Intangible assets and value capture in global value chains: The smartphone industry* (Vol. 41). WIPO.

Duhaylongsod, J. B., & De Giovanni, P. (2018). The impact of innovation strategies on the relationship between supplier integration and operational performance. *International Journal of Physical Distribution & Logistics Management*, *49*(2), 156–177. doi:10.1108/IJPDLM-09-2017-0269

Fairphone. (2016). *Closing the Loop: The garbage collectors of the mobile industry*. https://www.fairphone.com/en/2016/07/14/closing-the-loop-the-garbage-collectors-of-the-mobile-industry/

Fairphone. (2018a). *Researching the Best Available Methods for Electronics Recycling*. Available at: https://www.fairphone.com/en/project/researching-best-available-methods-electronics-recycling/

Fairphone. (2018b). *Reuse and Recycling*. Available at: https://www.fairphone.com/en/impact/reuse-recycle/

Fairphone. (2021a). *Impact report - A challenge to the industry*. Available at: https://www.fairphone.com/wp-content/uploads/2021/06/Fairphone_Impact-report_2020.pdf

Fairphone. (2021b). *Impact Report 2020: Fairphone nearly doubles sales in 2020 as its industry influence increases*. Available at: https://www.fairphone.com/wp-content/uploads/2021/06/Impact-Report-2020-Fairphone-nearly-doubles-sales-in-2020-as-its-industry-influence-increases.pdf

Fairphone. (2021c). *Fair Materials*. Available at: https://www.fairphone.com/en/impact/fair-materials/

Falcone, P. (2018). *Fairphone, un'azienda ed uno smartphone stile libero: ecologico, equo, solidale e sostenibile*. Io Comunico. Available at: https://madeingaia.it/it/prodotti-sostenibili-persone/per-le-persone/comunichiamo/25-fairphone-un-azienda-ed-uno-smartphone-stile-libero-ecologico-equo-solidale-e-sostenibile

Jalali, H., Ansaripoor, A. H., & De Giovanni, P. (2020). Closed-loop supply chains with complementary products. *International Journal of Production Economics*, *229*, 107757. doi:10.1016/j.ijpe.2020.107757

Kitchin, R. (2018). The realtimeness of smart cities. *TECNOSCIENZA. Italian Journal of Science & Technology Studies*, *8*(2), 19–42.

Kortmann, S., & Piller, F. (2016). Open business models and closed-loop value chains: Redefining the firm-consumer relationship. *California Management Review*, *58*(3), 88–108. doi:10.1525/cmr.2016.58.3.88

Kumar, V., & Dahiya, B. (2017). Smart economy in smart cities. In *Smart economy in smart cities* (pp. 3–76). Springer. doi:10.1007/978-981-10-1610-3_1

Maranesi, C., & De Giovanni, P. (2020). Modern circular economy: Corporate strategy, supply chain, and industrial symbiosis. *Sustainability*, *12*(22), 9383. doi:10.3390u12229383

Nussholz, J. L. (2018). A circular business model mapping tool for creating value from prolonged product lifetime and closed material loops. *Journal of Cleaner Production*, *197*, 185–194. doi:10.1016/j.jclepro.2018.06.112

Reuter, M. A., van Schaik, A., & Ballester, M. (2018). Limits of the circular economy: Fairphone modular design pushing the limits. *World Metall.–. Erzmetall*, *71*(2), 68–79.

Sacco, A., & De Giovanni, P. (2019). Channel coordination with a manufacturer controlling the price and the effect of competition. *Journal of Business Research*, *96*, 97–114. doi:10.1016/j.jbusres.2018.09.001

Vishkaei, B. M., Fathi, M., Khakifirooz, M., & De Giovanni, P. (2021). Bi-objective optimization for customers' satisfaction improvement in a Public Bicycle Sharing System. *Computers & Industrial Engineering*, *161*, 107587. doi:10.1016/j.cie.2021.107587

Wernink, T., & Strahl, C. (2015). Fairphone: Sustainability from the inside-out and outside-. In *Sustainable value chain management* (pp. 123–139). Springer. doi:10.1007/978-3-319-12142-0_3

Chapter 5
Circular Economy at the Core of Levis & Co.'s Success:
The Circular Business of Denim

Riccardo De Lucia
IGT Lottery S.p.A., Italy

Armando Iavazzi
Luiss University, Italy

Ilaria Combatti
Aglaia Capital Partners, Italy

Benedicta Quarcoo
IFAD, Ghana

EXECUTIVE SUMMARY

The chapter aims to gather information and analyze the new business model developed by Levi Strauss and Co. First, it analyses the network that LS&Co. has set up and how the business works. Then, it will investigate the context in which LS&Co. operates by looking at values, drivers, stakeholders, barriers. Subsequently, a careful analysis of the performance indicators that LS&Co. has adopted including the GRIs material report will be conducted. The technologies and developments that LS&Co. has adopted in its business will be explained; particular attention will be given to emerging partners in the new business model and the value they bring. There will be space for an ambitious project based on blockchain technology as a tangible tool for their products and suppliers. Some recommendations on how LS&Co. could improve its circular economy in the context of the growing smart cities will be given in the last paragraph.

DOI: 10.4018/978-1-6684-5001-7.ch005

INTRODUCTION

The secular adoption of a linear production system adopted by almost all the companies (resource, manufacturing, use, disposal, and incineration) (Kenniskaarten, 2013) in the economic environment has demonstrated its large limits and boundaries. The unsustainable carbon footprint and waste of value are no longer socially accepted and harmful.

This social movement has led many companies to adapt their business to the new value drivers required (Economic benefits, pollution risk, resource recovery, waste management), adopting new technologies and business models to remain competitive in the market. Some of these companies such as LS&Co. were far-sighted and adopted these changes early on. Today, LS&Co. is among the most innovative companies in the apparel industry and the most involved in the circular economy, its involvement is well documented since 2015 (Duhaylongsod & De Giovanni, 2018).

The fashion industry is among the most impactful industries globally. The social impact makes it the focus of major issues with employment levels close to 1.7 million and an industry worth 512 billion only in the EU area (Fashion United, 2016). While the environmental impact is the cost that must be paid for such a large industry, much of the materials used are plastic and the useful life of the product has been reduced over the years. Fast fashion brands are among those most responsible because of the business model adopted, with a new collection every week. Specifically, given the low cost of production, the most used material in the low-price bracket of the market is polyester, which has overtaken cotton in global demand(Rinnovabili e Risparmio, 2019). The use of polyester material exposes the risk of dispersion of micro-fibers into the environment during washing.

Based on a McKinsey report (Berg et al.,2020)) the fashion industry is responsible for 2.1 billion tons of CO_2 emissions in 2018 with a CAGR of 2%, while the water consumption currently is around 93 billion cubic meters per year (Social Share, 2021). In addition, the fast fashion industry has changed consumer behavior, leading consumers to buy more frequently throughout the year and to wear less clothing, thus reducing the time of use and bringing closer the end-of-use. These 4 main underlined factors are the identified general concern and priorities that LS&Co. has identified for structuring its Circular economy system.

LS&Co. has approached each of these issues individually and with different activities. LS&Co. has implemented operations for the recovery of resources and the use of renewable inputs against CO2 emissions and water consumption, with the objective of carbon neutrality. The Investments in new technologies and joint ventures, such as the FLX project for fiber reuse. Finally, operations related to product return, i.e. recovery and extension of product life through repair combined with strong marketing activities to extend product life and change consumer behavior (Vishkaei et al, 2021).

CIRCULAR ECONOMY NETWORK

The fashion industry is the second-largest polluter in the world after the oil industry and its environmental damage is increasing as the industry grows. It has been shown that the average life span of a garment depends on the number of times it has been worn and can go up to ten times, creating a huge impact on natural resources (De Giovanni and Zaccour, 2022). Levi's, a giant in the apparel industry for over 168 years, has worked in recent years to move the apparel industry towards a circular future, where materials are used and reused safely. This allows making products with their next use in mind, using the "product for X" methodology.

LS&Co. has always taken direct action to protect the environment, from technology innovations for long-lasting products to new production processes such as Water-Less, that reduce the water consumption of jeans, or cottonized hemp. They have initiated and implemented great ideas that reduce the impact of products and help promote sustainable clothing (Maranesi and De Giovanni, 2020).

On the other hand, used or existing products, the famous blue jeans, are introduced into the circular process in different ways. It can be noted that Chip Bergh, CEO of LS&Co., openly stated at Fortune's Brainstorm Green Conference: 'stop washing your jeans'(2014). Levi's CEO dwelt on this because according to an LCA (life cycle assessment) esteem of a pair of jeans, conducted in 2013, half of the water consumption occurs at home during washing. Another, LCA conducted in 2007 by consumers found that washing products every 10 uses instead of every 2 reduce water intake by up to 80% (De Giovanni, 2016).

Bergh says that LS&Co. is committed to making "jeans last longer than the average person's waste lines". To help increase the lifecycle of jeans, LS&Co. has begun working with several companies including Re:Newcell and its Circulose (De Giovanni, 2019).

Figure 1. LS&Co. Circular open and closed loop. Source: Incollect GMBH (2019)

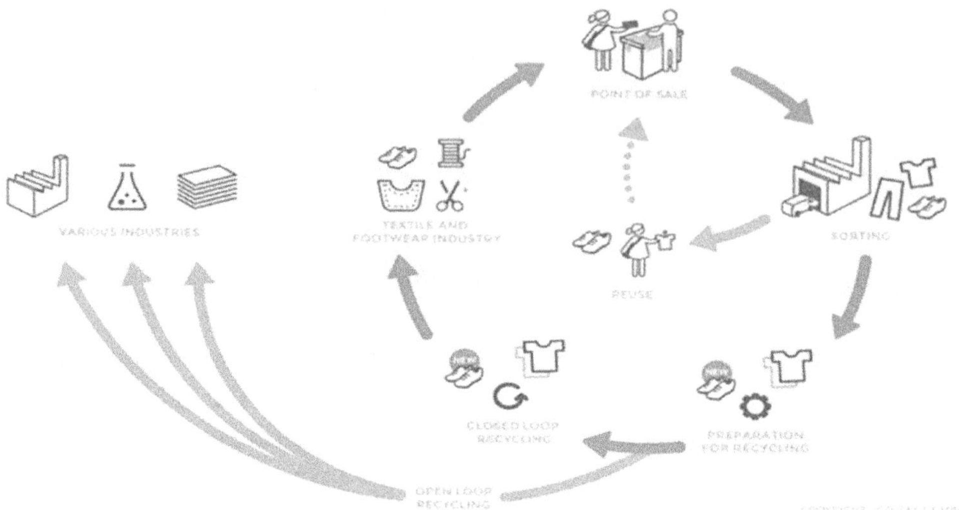

LS&Co.'s goal was to create an infrastructure from 2015 that supports closed-loop products by 2025 and recycling old LS&Co. products. Thus, reducing the impact of cotton farming by harvesting denim from people's wardrobes that would otherwise end up in landfills (De Giovanni, 2020, 2022).

To bring the circular economy vision to life, LS&Co. aims to increase research and development efforts to find ways for separating denim made from cotton and polyester blends so that these fibers can be recycled (Jalali et al., 2020). A further objective is to work with suppliers to research and develop cotton fibers that are strong enough to be recycled and still meet high-quality standards. In addition, LS&Co. has opened tailoring shops worldwide that can provide services like repairing, resizing, and refurbishing worn types of denim. This action is another important step towards achieving a circular economy.

Circular products also reduce risks throughout the supply chain by using more sustainable fibers circular products reduce companies' vulnerability to environmental risks and reduce the expense for clothing waste disposal in landfills, while also meeting growing consumer demand for more sustainable and circular clothing options. In doing so, LS&Co. can preserve customer value while extending product durability and usability.

LS&Co's circular economy implementation process starts in the brand's shops, where consumers are encouraged to bring back their used clothes by placing them in collection bins, in exchange for a 20% discount on brand-new denim from their sustainable line-up. From a marketing point of view, they are pushing sales of a more expensive brand line without dispersing its intrinsic value. The discount works even if a customer brings a different brand of jeans, in this case, LS&Co. considers that as a benefit because it is not raw material that they produce and for which they face initial costs. The clothes are then collected, sorted according to quality and future use because the technology that LS&Co.is using now allows recovering the denim based on the fiber's quality. The first option is recycling it, using Re:Newcell technology, so transforming the fibers in Circulose for making new types of denim. The second option is actionable for the most worn jeans that cannot pass through the recycling process; they are shredded and used for insulation materials in-house solutions (De Giovanni, & Vinzi, 2014).

The collected products are transported to the sorting centers, where the materials are identified and separated using automatic identification methods. Part of the clothes considered as not used at all will be stored and sold as second-hand goods, so that they can be reused rather than recycled, providing additional economic compensation for the company. The batch part that cannot be worn will be put into the closed circuit and recycled to create new clothes, by recovering the fibers and the small metal parts. The recovered fibers will be integrated into the partners' supply chains, in a closed-loop recycling approach, sent to Re:Newcell or they will be reused as materials in various sectors, most of them are turned into insulation material for the industry, by joining an open-loop recycling approach.

LS&Co. recently allows collecting very problematic products for final consumers, footwear. They recognized that many users are not aware of how to dispose of these kinds of things; LS&Co. decided to test the collection and recycling process without structuring a proper circular loop. In the case of footwear, the rubber granules resulting from recycling will be used to create new soles for shoes, joining the future planned closed loop. Alternatively, through a specialized shoe recycling plant, that mechanically dismantles shoes into their components and obtain usable secondary raw materials. The materials received from recycling will be then used to manufacture new products for other sectors, another open-loop recycling.

Finally, LS&Co. has partnered with Maersk which is a global logistic partner which has an ECO delivery program that uses waste-based biofuels. LS&Co. has been a member of the Cargo Working Group since 2018, they are a consortium of eco shippers-access to emissions data. Maersk is a big player in the logistic sector with a well-structured network thanks to its membership into LEO Coalition which pursues environmental benefits and commercial viability that supports sustainable shipping.

Analysis of the Value, Drivers, Stakeholders, and Barriers

Regarding the values, since its founding, LS&Co. has been one of the most committed U.S. companies to address social issues, as they state, "always stand up for what's right". They reject racial segregation, encourage LGBTQ rights, and ensure worker protection in their worldwide supply chain through corporate integrity and transparency. Starting from 2015, LS&Co. considers circular economy objectives for future generations and has settled their Circular Economy values. LS&Co. sustainability program is recognized for its innovative strategies in a wide range of fields such as climate change, water use, chemistry, and the use of fibers for enhancing product innovations. For example, the company is committed to the use of Cottonized Hemp and denim recycling technology. In addition, LS&Co. returned to profitability and sales growth from 2019 thanks to the management brand enhancement and a new recyclable line-up that give new life to their iconic product portfolio. Finally, is possible to synthesize LS&Co. CE values into a precise renewal long-term plan, to align the economic, operational, and social sustainability that has always distinguished the company's brands with environmental sustainability.

Regarding the drivers, LS&Co. has an ultra-centenary history, and its corporate drivers are in continuous development, focusing on integrity, performance, sustainability, durability. These corporate drivers guide their choices in its circular economy strategy, leading LS&Co. in new technology adoptions like their Water<Less® productions and Tencel™ Lyocell fiber.

About the stakeholders (Levi Strauss & Co.,2020), LS&Co. is very committed in its value drivers, one of which is transparency, which is why they have made explicit who they consider being their stakeholders and with whom they remain in contact to understand their perspectives on key issues such as environmental, social and governance issues. For each stakeholder, they have also given the manner of engagement; for consumers, they have mainly touched into the LS&Co. stores or by email. They are touched also by creating engagement on social media and participation in the LS&Co.® Red Tab™ loyalty program. Regarding the environmental and sustainability topic, they frequently use newsletters and information on the e-commerce site about product sustainability attributes. The company also uses surveys on environmental concerns and perceptions of the brands' sustainability commitments. Focusing on input and collaboration policy development in their annual sustainability summit for licensees, they increase employee commitment by using Surveys and Team meetings. The Company intranet is a useful tool to update the employees by using emails from Leadership, Employee Resource Groups, and Ethics and Compliance Report line.

The shareholders and prospective shareholders are main players in the LS&Co. sustainability route and to them, the firm dedicates one-on-one engagement, conferences, and annual formal communications including their press releases and forms. LS&Co. is monitoring their suppliers and supply chain workers because they are aware of their fundamental role in the business change and for this reason, they enhance workers' well-being. The supply chain workers are involved in studies and surveys for suppliers' assessments. Even collaborating with the "SHINE" initiatives at Harvard University. Finally, the suppliers are subjected to constant feedbacks and interviews during workshops and training. Also, NGOs, brands, and industry coalitions are invited to meetings, live events, industry group roundtables, partner collaborations, and surveys. Key Memberships and Partnerships are considered key for success by LS&Co. Within the most relevant organizations engaged, they have Accounting for Sustainability, BSR, Canopy Planet, Ellen MacArthur Foundation: Jeans Redesign Project, UN Global Compact, and the Water Resilience Coalition. As it is possible to see, they are all involved in changing fashion business issues as raw material consumption, water consumption, and climate change.

The barriers can be generalized to all fashion industries involved in the sustainable and green economy conversion. There are 4 key barriers:

1. Supply chain (Bioregional,2020) is the first large barrier. Fashion retailers work in a crowded, highly competitive environment and so they tend to structure a complex web of suppliers in different geographical positions. Considering that, dealing with a Circular economy implies the use of a suitable feedstock, and several suppliers must change their way of work, or the firms need to find new suppliers with high risks. In the LS&Co. case they had to deal with all 4 geographical areas of their production networks.
2. The Consumer attitude (Zhang et al.,2021): Over the years, consumers have been subjected to making purchases for each collection adapting to a fashion system based on at least four (4) collections per year. As already known, many items of clothing remain unused and never reach the end of their life. To implement a circular economy system, companies

need to establish an active approach to product return by providing an incentive for the customer to return unused clothes. LS&Co. approach that barrier by encouraging the consumer to simultaneously return and buy back a new garment. In the specific, that garment is produced with recycled resources that other customers have already returned to the company. This is a means of providing new resources and maintaining marginalization in its closed loop.

3. Higher final prices (Ivanova I.,2019): Setting up a sustainable business model means first achieving economic sustainability. The company can only finance investments that provide future positive cash flows; while a closed-loop often involves more steps in processing, know-how, and additional partners which entails higher costs. This barrier may be non-existent for luxury brands that make their price strong for brands such as LS&Co. whereas implementing uncompetitive prices on the market means reducing their sales volumes. Many companies have compensated for this price increase through careful and transparent marketing planning (Perrault M., 2021). The transparent pricing approach helps the consumers to feel involved in the environmental cause by clearly conveying why there are two prices for nearly the same garment. The company justifies to the customers because the same item has two different prices and that, is not just "Marketing". LS&Co. has a lot to do on that side apart from the discount they currently provide on product returns from a marketing point of view. LS&Co. does not justify the price difference between its circular and linear products. It must be underlined that there are some exceptions in the line-up, like the WaterLess products not properly "Circular", which have pricing almost identical to the "standard" one.
4. Quality, durability, and recyclability (Business of Fashion, 2021) of products are a challenge for all apparel firms. Quality is very tricky as an argument because it depends on the technology and the materials used for making the goods. LS&Co. is very focused on reaching its operational sustainability in its Circular business model by investing in innovative technology. About recyclability, it is a major concern because the company must choose very carefully the materials they want to recollect and their second or third life use. After all, not all recycled raw materials can be recovered or refurbished to the initial state and performance. LS&Co. has developed a specific partnership, indeed they recognize the limits of their denim is being re-constructed and decided to use that material for making insulation materials.

How the Circular Economy Impacts the Corporate Strategy

LS&Co. like many other companies has always had a linear economic model contributing to waste generation. In 2015, within clothing waste, only 15% has been recycled. The company had fully embraced and implemented the concept of circular economy in its corporate strategy. They have implemented the possibility of bringing old clothes to any LS&Co. store in the U.S. and they will collect them and reuse or recycle them with its partner I: CO, which provides the know-how for recycling and reuse in the textile industry. As stated by LS&Co., this action will avoid 24 billion pounds of clothes from going into landfills by transforming old jeans into insulation materials, cushioning, and new fibers. This requires huge investments in R&D, specific infrastructure, and corporate changes that support a circular economy model. In this case, LS&Co. had to modify the existing plant for the separation of denim made of cotton from polyester, changing its governance and monitoring strategy. Then structure the new logistics with its new partner.

The reason behind some LS&Co. choices is the product lifetime value which is at the center of the company proposition. LS&Co. has more than 80 shops in which a customer can bring their denim for repairing, resizing, and restyling so that the life of a product can be extended. In addition to that, the company is also collaborating with "Sustainable Apparel Coalition" to better design, produce and recycle denim embracing the concept of "Design for X" (Tsai-C et al., 2001). All these initiatives require a huge amount of effort and increase operational costs from LS&Co. In the case of partnerships, it would require time and money to build the right contract but it would positively impact the corporate strategy since now more customers are concerned about a company's sustainability commitment.

In 2020, the company launched different initiatives to use sustainable fibers using cottoned hemp and pushed towards a cultivation method that requires less water and fewer pesticides. They aimed at reaching 100% sustainably sourced cotton by 2025 (currently 83%). Since there have been challenges to the global supply of organic cotton, LS&Co. wants to not only increase its use but also increase the traceability of its sustainable agriculture. Within the initiatives, there is the exclusive use of organic cotton, thus "the better cotton initiative" to enhance sustainable agricultural practices, the "U.S. Cotton Trust Protocol to set higher standards for cotton, the use of cottonized hemp, and the use of pioneering fibers. Those initiatives have a relevant impact on

Figure 2. Jeans Life Extension Scheme. Source: Levi's Strauss (2019)

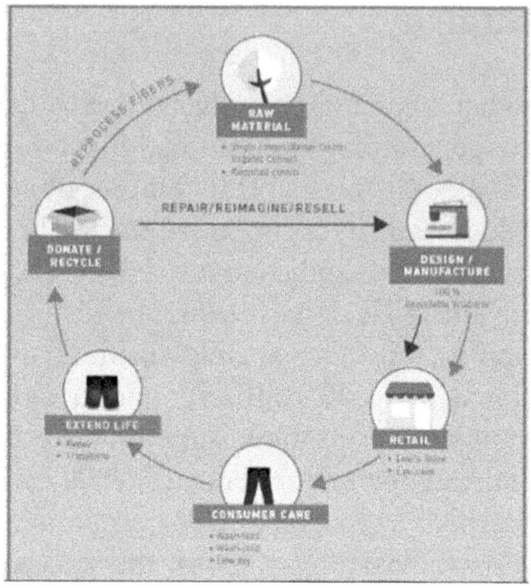

its corporate strategy affecting: the supply chain and logistics, the internal organization, and overall go-to-market operations.

Regarding the supply chain, the company requires its suppliers to comply with a series of social and environmental requirements codified in the "Terms of Engagement". Some of these requirements regard gender equality, wages, health, safety, and migrant worker protection. As stated by its sustainability report also makes sure that its suppliers contribute to addressing key social and environmental issues. The company also applies a sustainability materiality assessment to assess emerging issues to the business and its stakeholders. All those constraints make it difficult for LS&Co. to find the right suppliers considering the spread of their geographical production and the large volume of materials required. Many suppliers do not meet their requirements especially on worker wealth and the sustainability KPI. Therefore LS&Co. opts for building long-lasting and well-constructed contracts when finding out the right one and to monitor free-riding behavior, they have also invested in blockchain technology to help increase transparency.

The internal organization has changed in LS&Co. since they introduced a Global Sustainability Team for the creation and execution of a sustainability and circular strategy across the global LS&Co. company. The team is composed of heterogeneous members and works with functional executives to ensure the application of actions. The company approach is that a challenging environment with "management by objectives" is dependent on the company's sustainability results and therefore executives will be eligible for compensation. In the end, LS&Co's, corporate strategy is a hybrid strategy as stated by Doyran (2020), focuses on cost leadership and best-price value. The main efforts have gone towards cost reduction in the supply chain, marketing to improve women's wear, product innovation, and diversification. As reported by Preece, Fleisher, and Toccaceli (1995), LS&Co. uses its corporate strategy to build a positive global reputation to create a competitive advantage over its competitors.

About the overall go-to-market operations in 2020, LS&Co. sharpened actions on areas that were making an outstanding contribution to business performance. LS&Co. has a leading strategy with its brand called "direct-to-consumer first" while diversifying the brand portfolio by driving growth in high potential areas and digitally innovating the business. The company focused on the wasted values of its linear strategy by recovering the resources and the embedded value of its items. The wasted resources were mainly water and chemicals. Chemicals for coloring their denim while water consumption is strictly related to the production process and cotton-growing. For this reason, they recover those values using chemical protocols and the patented WaterLess technology in their plants.

The wasted embedded values are recovered by repairing, recycling, and upcycling approach. Recycling the denim that cannot be repaired and using the fiber to be reprocessed for making higher-value products like insulation materials. For doing so, LS&Co. had to change its marketing and customer communication- they use an active approach to clients and product return by giving a discount (20%), and then they nudge clients to purchase the new circular products. In the end LS&Co. has strengthened its brand heritage and connections with consumers through product purchasing.

LS&Co. is very concerned about the future evolution of its business, and therefore it has already expressed the next strategies and evolutions of its circular economy. The management wants to make the entire LS&Co. production circular and carbon-neutral, anticipating its competitors. They are investing in the "jeans redesign project" (Levi Strauss & Co., 2021, Designing a more circular future), a more concrete application of the already stated concept of "Design for X", and the aim is to create a production standard for jeans that is more durable and easier to repair and recycle which will be replaced soon by the classic structure of LS&Co. jeans. LS&Co. produces a wide variety of products adopting also synthetic fibers and for this reason, in the next years, it plans to implement a circular loop also for other types of products such as synthetic jackets and shoes. They will use the same drivers that led them to develop such a well-implemented loop for its iconic denim combining the know-how acquired by its partners in these years.

All the future corporate strategies of LS&Co. will be direct to increase marginalization on such products for which LS&Co. faces considerable costs for recovery and treatment. LS&Co. pursues higher margins by exploiting the recovered feedstocks for making denim and by not selling fiber to partners. Thus, maintaining the value inside the company and lowering the costs of virgin materials.

The Links Between Circular Economy System and Performance Indicators

LS&Co. annually publishes its sustainability report (Levi Strauss & Co., 2021, Sustainability Report) in which its own key goals, performance indicators, and the progress reached through the year are clearly explained. The report is structured into climate, consumption, and community. For a clearer analysis, it is quite useful to consider the consumption part of this report to get useful insights about LS&Co's circular economy performance.

Regarding the recovering of resources such as water, LS&Co. declares that 70% of its denim is part of the Water<Less program. The adoption of this program has managed to save 4.2 billion liters of water since its introduction as well as saving on energy since it is moving towards 100% use of renewable energy.

LS&Co. explains its sustainable inputs usage and the company uses 75% of the cotton sourced through the "Better Cotton Initiative" while aiming to reach 100% (Levi Strauss & Co.,2021, SASB Index) man-made cellulosic fibers. 80% of chemistry suppliers are implementing the "Screened Chemistry" scheme, which covers 1200 chemicals and has led to the elimination of 16 dangerous chemicals from the manufacturing process. Their jeans are made with 55% hemp-blended using natural dye and post-consumer recycled fibers, more organic cotton, and the recyclability of every "WellThread" garment.

2021 is the first year of use for their cottonized hemp with a production of 12,000 units but they have scaled this innovation and the forecast is to deliver more than a million units in less than two years. Looking at the LS&Co. GRI (Levi Strauss & Co., 2021, GRI) in the material parts 301,302, and 303, a deeper explanation about all their sustainable materials, practices, and indicators are highlighted. 83% of the cotton (Levi Strauss & Co., 2019, Sustainability Review) comes from sustainable resources thanks to all the activities already reported, then 95% of post-consumer waste is used for making retail paper bags.

About the recycling of denim fibers, the technology (Levi Strauss & Co., 2014, How recycled jeans can keep you warm) to transform jeans into insulation material is already known in the market and allows the fibers to be disassembled and 80% of them are recovered as new feedstock. While using the Re:Newcell technology, LS&Co. can turn 100% of recovered fibers into Circulose® and use it for making new clothes by substituting cotton fibers.

For the collection of used jeans, LS&Co. has seen an increase in the return of used consumer products, thanks to its "Jeans Go Green" program in the US and they collected 132.399 discarded jeans in 2019, saving 66 tons of waste that became insulation materials. Considering the 2019 annual jeans sales is a very small number, but it is a starting point. LS&Co. aims to reach 100% of renewable production but avoids declaiming the percentage of products that return from customers.

About the economic performance- sales of second-hand jeans have started in 2020. In the period of October-November 2020, LS&Co. sold 1.132 pairs, with a total of units posted on the "SecondHand" site in 2020 was 4799. This is a small amount, but the environmental impact is huge since to produce a new pair of jeans, ten thousand liters of water is required. Overall, company performance is quite good, apart from 2020 (the pandemic year). Net revenues, earnings per share, EBIT, and gross margins have all constantly increased during the latest years. LS&Co. managed to open new shops in 2020 bringing the total number of flagship stores to 1.042.

As stated earlier, they do not exhibit the percentage of their own recovered denim, but they are working in the future to implement a tracking system that will enable them to determine that percentage. The company is investing in digital and artificial intelligence tools to empower workers and to increase the efficiency of the global supply chain, optimizing promotions and assortment planning. Globally, its circular systems helped the company in improving the performance indicators thanks to cross-sectional benefits like the brand enhancement of their sustainable marketing campaign and activities.

Circular Economy Developments and Business Transformation

LS&Co is an 1853-founded company based in San Francisco. Being a firm with more than 100 years of operation, its business and product portfolio has evolved over the decades. The production of the initial and iconic jeans has been joined over time by a wide variety of products such as jackets, shirts, t-shirts, and shoes. The predominant materials used by LS&Co. are polyester, leather, and cotton. The company has always been able to adapt and anticipate the trends or needs of the environment in which it operates, and it is one of the only two major companies in the textile industry with commitments under climate goals stated in the Paris Agreement.

LS&Co. has identified the goals of carbon neutrality and the adoption of new, more efficient production methods as the next step in the evolution of its business model. The business transformation has implied the adoption of a circular business model combined with a series of activities related to the usage of renewable energies and the avoidance of wasting resources such as water.

In 2007, LS&Co. conducted the LCA (Lifecycle Assessment) to assess the lifecycle impact of its core denim products, the Levi's 501. The LCA demonstrate that cotton was the most impactful material in the use of non-recoverable primary resources: water and energy. They also detected a correlation to specific phases of the value chain, production, and post-purchase care.

LS&Co. structured a partnership with Goodwill® joining the "Care Tag for the Planet initiative" that nudges consumers to learn new behaviors in post-purchase care. The objective was to take down energy and water during the product lifecycle. Regarding the cotton agriculture impact since 2010, LS&Co. joined "The Better Cotton Initiative" to invest in reducing fossil-fuel-based chemicals in global cotton agriculture. They then combined with the U.S "Cotton Trust Protocol" to reduce synthetic chemicals for better soil health which in turn can absorb more carbon dioxide.

LS&Co. has decided to renew the raw materials it uses for the products it is committed to, using only organic cotton and where necessary only recycled polyester. The aim is to reduce the carbon and climate impact of their materials combined with the use of the renewable source of energy which will lead LS&Co. to achieve its emission reduction targets by 2025. The 2025 emission reduction target of LS&Co. (Levi Strauss & Co., 2018, Climate Action Strategy) is based on a specific climate protection strategy planned in 2016 with different objectives across all direct and indirect operations. They use a science-based approach to target emission goals in different points of their value chain. For owned and operated facilities, they esteem to move from 90% to 100% renewable electricity and reduce emissions by 40% across the supply chain. The consumer care commitment does not have a target expressed in numerical terms as it is difficult to track consumer use. However, LS&Co. is committed to driving a new way of using products throughout their usage as they state on their care label "wash less, wash cold, line dry, donate or recycle".

LS&Co. circular economy loop is aligned with the imposed climate targets. The technology "Water<Less®" used in the production of the jeans since 2011 minimizes the use of water, maintaining the same quality and resistance characteristics of standard production. The jeans were not re-used to make new LS&Co. jeans, but it was reprocessed and shredded to be sold as insulation materials for construction. This solution, however, is an option that leads to a downcycling of the product as the recovered value is much lower than the intrinsic value of LS&Co. jeans.

In 2020, thanks to its new partner Re:Newcell, they adopted "Circulose" that solves the problem of reusing textile fibers enabling LS&Co. to recycle its jeans for making a new one, thereby recovering all the intrinsic value of denim. In the end LS&Co. will recycle good condition fibers in new denim jeans while the worst part of the collected jeans is going to be used as a lower value product for insulation.

They realized a new line-up consisting of waterless models with the same profitability as the old classic jeans and those made of "Circulose" which can be purchased by giving back an old pair of jeans taking some margins by the collected denim. The store becomes the collection point located around the world, managing the return logistics through the standard distribution following it backward.

For the next step, LS&Co. is planning new initiatives to recycle its synthetic fibers, mainly polyester and nylon, exploiting the acquired know-how in the logistics and contracts for partnerships with third entities.

LEVI STRAUSS BLOCKCHAIN TRACKING FOR RAW MATERIALS AND MARKETING IMPLICATIONS FOR CIRCULAR ECONOMY

LS&Co. developed and use a blockchain to track worker well-being at factories that supply Levi Strauss. The system anonymously and securely tracks worker well-being on a blockchain and the benefit is that once the data is recorded, it becomes tough to tamper with it. LS&Co. can therefore use the blockchain strategy to enhance the circular economy by using blockchain to turn the circular economy into a developed ecosystem by offering product assurance and active incentivization. Blockchain has an enormous potential of positively impacting LS&Co. business solutions, practices, and performance with the circular economy in this regard. In addition, the marketing implications of a blockchain are numerous and thus, blockchain technology has already been adopted in other businesses as in the case of Cantina Placido Volpone. Consumers are more conscientious about what they buy if they know where a product comes from. If they have more certainty about its sustainability, consumers are willing to pay more, but certifications are not enough. The blockchain is therefore a starting point for developing LS&Co's. own direct marketing touchpoint with customers, knowing their habits, and also their use. LS&Co. can then trace the path of the jeans even in the aftermarket and once collected for recycling can estimate the average time its consumers use the jeans, and thus better structure the resources needed in its circular loop.

Another emerging technology by LS&Co. is the F.L.X (future-led technology) to bring back vintage pairs of jeans using the most contemporary technology around, all while reducing water wastage, combating overproduction, and making sure it can react to trends almost immediately. The technology recreates the looks and feels of vintage denim. The main purpose of the FLX technology is to change customer behavior and approach to their old garments making customers think about what they buy, how they wear them, and what to do after wearing them completely. A way of creating awareness indirectly to enhance the circular economy.

Other Smart Cities Recommendation to Levi's

The smart collecting point would be the next step. Although LS&Co. already has a way of collecting and recycling used garments, LS&Co. can ameliorate its recycling of textile system by partnering with Re: Newcell. This project would be perfectly fitted to new, upcoming initiatives in smart cities like Zero Panik and the aim is to structure and promote the collection of clothes within cities, buildings, and households managing waste more efficiently. The Zero Panik initiative is a strategic means for LS&Co. to look upon, specifically for collecting used garments both online and offline. An example is the use of textile recycling smart bins placed at vantage points for collection, and Levi's can add a part of this strategy to theirs to get a holistic system of collection, tracking, incentivizing, and recycling.

LS&Co can learn from Dandong City in China that is developing the new concept of the textile and clothing industry through 5G. Building intelligent workshops and intelligent factories, pushing acceleration in the construction of new manufacturing systems through collaborative research and development of intelligent production. In the end by avoiding waste of resources, time and increasing inner productivity. LS&Co. can also implement the "Internet +" strategy, cultivate new formats and new models such as personalized customization and intelligent sales, enhance brand awareness and market share. Inject power into the textile and garment industry to accelerate its integration into the new development pattern cycle.

As part of LS&Co. processes to become smart, they can create a Levi's Cloth Collector app, specifically for collecting used clothing from their customers. It will be linked to sensors, IoT, and wireless connections to the bins, and anytime a garment is dropped, it will send a notification to the main Levi's collection system. The app will track the exact number of garments deposited through pick-up by the LS&Co. truck to disposal at the sorting center. All through these processes, the technologies connected will send and receive notifications. The system itself will award the customer according to the level of garment deposited.

CONCLUSION

LS&Co. demonstrates that the implementation of a circular economy in the textile industry can enhance sustainable growth and economic recovery within the planetary boundaries by restoring natural systems, reducing the negative impacts of climate change, and maintaining the minimum use of raw materials. (Lakatos et al., n.d., Conceptualizing Core Aspects on Circular Economy in Cities).

It is crucial to agree with the sixty-six companies that are part of the Ellen MacArthur Foundation's "Make Fashion Circular" initiative whose ambition is to introduce a system that keeps materials at their highest value while regenerating the environment. The Foundation further explained that in a circular economy, textile products are used more and are made to be made again, so they can be reused, remade, and ultimately recycled.

The other initiatives that Levi Strauss is taking to become an environmentally responsible company are all part of a global long-term strategy to remain a leader in a challenging market. The textile industry is one of the world's main polluters and attention to the sustainability of fast-fashion and the business model of sectors big corporation is strong among consumers. The indiscriminate exploitation of resources such as water, energy, and workers in Asian countries is damaging the companies. Even the regulatory power, especially in Europe, has already begun a process of reducing environmental impact as has already happened in the automotive sector.

In this context of change, LS&Co. is therefore in an advantageous position having worked over the years to build its supply chain, reduce resource waste combined with high social standards and ultimately build its circular economy. By analyzing the European agenda 2050 for the circular economy, LS&Co. seeks to be aligned with the objectives of climate neutrality.

The managerial implications emerging from LS&Co's new circular business are very interesting. The path takes time, time to renew a linear production system into a circular one. The circular economy is not easily attainable, but it needs constant investment over time, and the management must carefully plan the way forward. The technologies are not all reliable, entrusting the entire production to process technology that is not entirely adequate can undermine the economic and operational sustainability of the company. LS&Co. has mapped out the way for the others with a process that includes a holistic view of the company's business. It is about expanding product portfolio starting from iconic and recognizable elements, testing already available technologies on small volume productions to complement the core business. The circular economy has to be implemented on a small scale while the company has to be prepared to face the increased demand if the closed loop works properly. Choosing the right partners who can support the size of the business would be one of the key success factors. The process must therefore be gradual, the products must become sustainable with time to guarantee the company's cash flow and operations. Finally, building trust and teaching consumers a new way of purchase and that the active approach to return is key to a successful circular loop in the textile industry. Making the return of the old garment an integral part of the purchasing process, thus, becoming a necessary element of the purchase and recovering the value as it already happens for other consumer goods such as batteries.

Therefore, to conclude after a careful analysis of the company, it is possible to see that Levi Strauss is working on becoming a real sustainable company. These sustainable efforts towards building a circular economy will help LS&Co. enhance its brand reputation, attract stakeholder engagement, and make the company competitive leaving the environment liveable for all.

REFERENCES

Berg. (2020). *Fashion on climate*. https://www.mckinsey.com/industries/retail/our-insights/fashion-on-climate

Bioregional. (2020). https://www.bioregional.com/news-and-opinion/fashions-failings-barriers-to-a-sustainable-clothing-industry

Business of Fashion. (2021). https://www.businessoffashion.com/opinions/sustainability/a-more-circular-fashion-industry-will-require-a-collective-effort

De Giovanni, P. (2016). Coordination in a distribution channel with decisions on the nature of incentives and share-dependency on pricing. *The Journal of the Operational Research Society*, *67*(8), 1034–1049. doi:10.1057/jors.2015.118

De Giovanni, P. (2019). When feature-based production capabilities challenge operations: Drivers, moderators, and performance. *International Journal of Operations & Production Management*, *40*(2), 221–242. doi:10.1108/IJOPM-04-2019-0309

De Giovanni, P. (2020). *Dynamic quality models and games in digital supply chains*. Springer Nature.

De Giovanni, P. (2022). Leveraging the circular economy with a closed-loop supply chain and a reverse omnichannel using blockchain technology and incentives. *International Journal of Operations & Production Management*, *42*(7), 959–994. doi:10.1108/IJOPM-07-2021-0445

De Giovanni, P., & Vinzi, V. E. (2014). The benefits of a monitoring strategy for firms subject to the Emissions Trading System. *Transportation Research Part D, Transport and Environment*, *33*, 220–233. doi:10.1016/j.trd.2014.06.008

De Giovanni, P., & Zaccour, G. (2022). A selective survey of game-theoretic models of closed-loop supply chains. *Annals of Operations Research*, *314*(1), 1–40. doi:10.100710479-021-04483-5

Duhaylongsod, J. B., & De Giovanni, P. (2018). The impact of innovation strategies on the relationship between supplier integration and operational performance. *International Journal of Physical Distribution & Logistics Management*, *49*(2), 156–177. doi:10.1108/IJPDLM-09-2017-0269

FashionUnited. (2016). *Statistiche della moda Italia*. https://fashionunited.it/statistics/statistiche-moda-italia/

Global Recycling. (2021). https://global-recycling.info/archives/5931

Ivanova, I. (2019). *Buying "green" is too pricey for the average consumer.* https://www.cbsnews.com/news/buying-green-is-too-pricey-for-the-average-consumer/

Jalali, H., Ansaripoor, A. H., & De Giovanni, P. (2020). Closed-loop supply chains with complementary products. *International Journal of Production Economics*, *229*, 107757. doi:10.1016/j.ijpe.2020.107757

Kenniskaarten. (2013). *How is a circular economy different from a linear economy?* https://kenniskaarten.hetgroenebrein.nl/en/knowledge-map-circular-economy/how-is-a-circular-economy-different-from-a-linear-economy/

Kuo, T.-C., Huang, S. H., & Zhang, H.-C. (2001). Design for manufacture and design for 'X': Concepts, application and perspectives. *Computers & Industrial Engineering*, *41*(3), 241–260. doi:10.1016/S0360-8352(01)00045-6

Levi Strauss & Co. (2014). https://www.levistrauss.com/2014/03/13/denim-insulation-101-how-recycled-jeans-can-keep-you-warm/

Levi Strauss & Co. (2018). https://www.levistrauss.com/wp-content/uploads/2018/07/LSCO_Climate_Action_Strategy_2025.pdf

Levi Strauss & Co. (2020a). https://www.levistrauss.com/sustainability-report/our-reporting/our-stakeholders/

Levi Strauss & Co. (2020b). https://www.levistrauss.com/wp-content/uploads/2021/09/2020-LSCo.-GRI-SDG-UNGC-Index.pdf

Levi Strauss & Co. (2020c). https://www.levistrauss.com/wp-content/uploads/2020/07/LSCo.-2019-Sustainability-Review.pdf

Levi Strauss & Co. (2021a). https://www.levistrauss.com/2021/07/21/designing-a-more-circular-future/

Levi Strauss & Co. (2021b). https://www.levistrauss.com/sustainability-report/our-reporting/report-in-brief/

Levi Strauss & Co. (2021c). https://www.levistrauss.com/wp-content/uploads/2021/09/2020-LSCo.-SASB-Table.pdf

Maranesi, C., & De Giovanni, P. (2020). Modern circular economy: Corporate strategy, supply chain, and industrial symbiosis. *Sustainability, 12*(22), 9383. doi:10.3390u12229383

Perrault, M. (2021). *How Companies Can Overcome the Barriers to Sustainability in the Fashion Industry*. https://scholarworks.uark.edu/cgi/viewcontent.cgi?article=1057&context=finnuht

Placido Volpone. (2019). https://placidovolpone.it/blockchain-vini/

Preece. (1995). *Building a Reputation Along the Value Chain at Levi Strauss*. Academic Press.

Renewcell. (2020a). https://www.renewcell.com/en/circulose/

Renewcell. (2020b). https://www.renewcell.com/en/section/about-renewcell/

Rinnovabili e Risparmio. (2019). *Sostenibilità l'impatto ambientale del fast fashion*. https://www.rinnovabilierisparmio.it/2019/01/11/sostenibilita-l-impatto-ambientale-della-fast-fashion/

Sacco, A., & De Giovanni, P. (2019). Channel coordination with a manufacturer controlling the price and the effect of competition. *Journal of Business Research, 96*, 97–114. doi:10.1016/j.jbusres.2018.09.001

Social Share. (2021). https://www.commonobjective.co/article/the-issues-water

Vishkaei, B. M., Fathi, M., Khakifirooz, M., & De Giovanni, P. (2021). Bi-objective optimization for customers' satisfaction improvement in a Public Bicycle Sharing System. *Computers & Industrial Engineering, 161*, 107587. doi:10.1016/j.cie.2021.107587

Zero Panik. (2018). https://zeropanik.com/

Zhang, Y., & Zhou, P. (2021). Consumer Attitude towards Sustainability of Fast Fashion Products in the UK. *Sustainability, 13*(4), 1646. Advance online publication. doi:10.3390u13041646

Chapter 6
Acqua Minerale San Benedetto: Italian Pioneer for Circular Economy

Louis Alarcon
Luiss University, Italy

Valentina Arangiaro
Simest S.p.A., Italy

Adrian Fernandez Rabe
SICK, Italy

Felicián Elekes
Luiss University, Italy

Juliane Gallersdörfer
Deutsche Gesellschaft für Internationale Zusammenarbeit GmbH, Italy

Katharina Gramiller
Gesundheit Österreich GmbH, Italy

Gloria Rotundo
WINDTRE, Italy

Kristóf Tölgyesi
Luiss University, Italy

EXECUTIVE SUMMARY

Sustainability is an extremely urgent and universal concern; however, becoming sustainable doesn't happen overnight. It is not an immediate transition, and in order to really take sustainability to the next level, experience and know-how must be upgraded over time. San Benedetto is the perfect example of a company that developed a strong commitment to the environment since its origins. It has embraced the circular economy paradigm since it has shifted from the traditional linear economic model, which is based on a take-make-consume-throw away pattern toward a closed loop business model in which waste becomes an input in the production process. Over the years, San Benedetto has reached important achievements and prizes as far as environmental protection effort, proving to the world its responsibility and its proactive attitude over the matter, that in turn led to increase customer loyalty and displace the competition.

DOI: 10.4018/978-1-6684-5001-7.ch006

INTRODUCTION

Acqua Minerale San Benedetto S.p.A. is an Italian multinational controlled by the Zoppas family and based in the area of the Municipality of Scorzé (Venice province) that operates in the field of non-alcoholic beverages of which it is the national market leader. It makes and sells mineral water, carbonated and non-carbonated beverages across the Italian territory and abroad. It owns six production sites in Italy, namely Scorzè (Venice), Popoli (Pescara), Donato (Biella), Nepi (Viterbo), Viggianello (Potenza), Atella (Potenza). In addition, it controls two production facilities in Spain, one in Poland and one in Hungary. According to 2020 financial data, San Benedetto proves itself to be a resilient company since, thanks to its 10 worldwide production settlements and to its active commercial presence in over 100 counties, it has been able to achieve 703 million of consolidated turnover, a total of 2115 employees and 4,6 billion bottles of water produced yearly. Its competitive advantage is ensured by continuous product and process improvements allocated to the development stages of the entire non-alcoholic offering. The leading position as far as mineral water and soft drinks is also attributable to some fundamental company's pillars that build a strong and ongoing value proposition: tradition, innovation, corporate culture, sustainable expansion and constant focus on the consumer are the key ingredients for the success in the name of "Made in Italy".

According to Italy RepTrak 2022, an important annual study on companies' reputation, San Benedetto ranks 74th in the list of the 150 most reliable companies in Italy. This achievement is not surprising since for 5 years it had the highest corporate reputation in the mineral water, drink and refreshment sector. The model measures the perception of stakeholders (population, employees, media...) based on 7 rational dimensions: products and services, innovation, work environment, governance, social and environmental responsibility, leadership and performance. In this context, it must be highlighted the company's strong commitment with respect to social wellbeing and environmental sustainability, not casually the adoption of a sustainable approach is part of San Benedetto DNA.

History in Brief: The Road for Sustainable Development

It is interesting to retrace San Benedetto's evolution provided that in less than thirty years, it grew from a small local producer to a company with a national as well as international presence that has been able to displace domestic competition by leveraging on innovation and sustainability. San Benedetto S.p.A. foundation dates back to 1956 when it began its activity in Scorzè as a local water bottler using glass bottles and the water coming from an ancient spring of great quality from which it took the name. In 1971 the current management was focusing on three fundamental pillars: cost reduction, plant renewal and the exploration of an alternative to glass. This strategy laid the foundation of the 'one way packaging' approach that represented one of the most disruptive innovations since plastic bottles would have been introduced. The solution, however, was already discovered in America by the chemist Nathalien Wyeth: it was a thermoplastic resin, suitable for food and completely recyclable, without harmful emissions and resistant, the so-called PET, polyethylene terephthalate, later patented in 1973. The first bottle of PET was delivered by the company in 1980 for the sale of orange juice. From that point on, San Benedetto exploited deeply the new technological innovation that reshaped the packaging industry as a whole and opened to valuable opportunities the beverage sector. San Benedetto vertically integrated the process along the supply chain, self-producing its bottles and becoming the first Italian company to use them in different formats (2, 1.5,1 and ½ liter). In 1984 the internationalization process began: the company signed a franchising agreement with Cadbury Schweppes International and began to produce and distribute the related product in Italy. Similarly, in 1988 a partnership with PepsiCo was enforced. In this way it expanded its activity abroad and begins to commercialize and produce in Italy the range of products of Pepsi, Seven-Up and Schweppes. Follow an agreement of equal cooperation between San Benedetto and Danone and a partnership for the production and packaging of Coca-Cola products for the European market. In 1997 the company delocalized a production plant in Spain (Valencia) under the name of Agua Mineral San Benedetto located in a strategic position close to Fuente Primavera spring. As a consequence, San Benedetto Mineral Water became a Group. In 1998 the company registers Push and Pull top, a smart cap that was properly matching the new market need related to the modern "On the Go" consumption. In 2001, it expanded into Eastern Europe through joint ventures with Danone in Poland (Polska Woda) and in

Hungary (Magyarviz Kft). One year after the company acquired a stake in Santa Clara C.P.A., a firm in the Dominican Republic, for the bottling and commercialization of mineral water. Later, a second Spanish plant in Loja (Granada) was built named Parque La Presa S.A. "Easy compact" bottle was a successful item launched in the market in 2006 in one litre format in order to face the increasing demand of water consumption outdoor and produced compensating all CO2 emissions generated during the entire lifecycle. The target markets, in addition to Western Europe (Austria, Germany, Switzerland, France and the United Kingdom) and the Eastern European area (Hungary, Czech Republic, Bulgaria and Romania) are: USA, Canada, Russia, Japan, Australia, Israel and as far as Southeast Asia, it serves Singapore, Malaysia and Thailand. The Group has developed a strong ability to operate in highly diversified markets, learning how to handle knowledge transfer process and exporting its know-how where the market seems attractive. In fact, exports of water and secondly of tea cover a considerable percentage of the company's revenues (>32% of the group's turnover).

San Benedetto has been able to set up a well-diversified portfolio, thus it is defined as a Total Beverage Company because in addition to its proprietary brands of mineral water and soft drinks, it also produces and commercializes other products: iced tea, chamomiles, flavoured waters, fruit-based beverages, sport drinks, aperitifs, and tonic water. Hence, San Benedetto is a multi-specialist and multi-channel company capable of satisfying consumer's requirements and multiple tastes providing different solutions with a variety of packaging and formats.

Constant attention is paid to the environment indeed huge investments in innovation, in production process optimization and in know-how refinement led San Benedetto to become an Energy Saving Company, reorganizing all operating and product development activities towards eco-sustainability goals. Eco-sustainability is conceived as a business mission: in this purpose San Benedetto's philosophy relies on the idea of improving the quality of its factories by using electricity generated from renewable resources, reducing the weight of bottles, using less virgin plastic and more recycled one (RPET) namely designing a new generation of 'eco-friendly' bottles resulting in production efficiency by compensating CO2 emissions, lower consumption of raw materials and optimization of the disposal of waste; it represents a virtuous model for the refreshment sector that is encouraged to adopt EcoDesign bottles. As an illustration, in the last 10 years more than 15 million euros have been invested in San Benedetto Spain for R+D+I (Research+develop ment+Innovation)[3] that granted bottling technology improvements as well as logistic robustness.

CIRCULAR ECONOMY NETWORK

Regard for environmental concerns and the pursuit of energy efficiency have been from early on part of San Benedetto's company culture. It makes continuous efforts to refine its circular economy approach and it is constantly centring its measures on total decarbonisation of the production process. San Benedetto was among the first companies at the beginning of the 80s to introduce in Italy a beverage packaging made out of fully recyclable PET. Additionally, they have been able to reduce the use of PET by decreasing the overall weight of the bottles by 20% in 10 years. The company's purpose is to embrace an energy saving philosophy that involves reducing CO2 emissions at every point in the production chain. It was the very first company in Italy, who signed an agreement with the Italian Ministry of the Environment on a voluntary basis in 2009 to measure and assess the company's carbon footprint. This assessment was necessary to be able enact the aforementioned reduction policies, which resulted in the 1 litre Easy water bottle in 2010, a carbon neutral bottle through reduction and compensation of production carbon emissions (PETnology, 2021). A further reduction of GHG emissions has been achieved due to the deployment of renewable energy and to the installation of solar panels in their production plants. In this context, San Benedetto's Spanish subsidiary has already achieved 100 percent renewable energy usage in their production settlements. Finally, the introduction of the "Internet of Things" in the company process as of 2013 helps San Benedetto to keep track of their environmental impact and estimate how intervention projects would improve San Benedetto's sustainability performance and their production lines (Mango, 2020).

Another notable action is the "Network-Project", namely a plan that has envisaged the creation of a production facility network throughout the Italian peninsula: six different production sites have been established in distinct Italian regions, therefore shortening transport distances and consequently reducing the related logistic emissions, at the same time the maximum proximity to customers was ensured (Mango, 2020).

The introduction of RPET (recycled PET) in their production is part of the company's plan to reduce the environmental impact and to fully embrace a Circular Economy approach. Over the years San Benedetto has increased its engagement in plastic recycling process and it has also employed 100 percent recyclable aluminium for beverage cans. They implemented the first Italian PET collection scheme in 2011 and included RPET in their packaging starting from 2012. Furthermore, San Benedetto started the "New Life for PET" project in collaboration with the region Veneto, that was carried out from 2013 to 2015. The project was expected to set up collection containers at the main sale points all around Veneto region in order to encourage a new environmentally conscious behaviour among customers. Through the initiative, customers were given the possibility to hand their used PET bottles, facilitating the collection and the recovery of PET. The material collected from post-consumer waste can then be used as RPET for the generation of new bottles or other products. This project was also done in cooperation with the plastic recycling company Aliplast S.p.A. who took care of turning the used PET-bottles accrued from collection into RPET for further use in the new bottles for San Benedetto (Mango, 2020). This project exemplifies the needed collaboration between customers and players along the supply chain for an efficient Circular model. What also stands out is the commitment of the company in waste management, namely the continuous attention also to post-sales activities.

Analysis of the Value, Drivers, Stakeholders, and Barriers

The main stakeholder participating in the ambitious sustainability scheme of San Benedetto is clearly the Italian Ministry of Environment. In 2008, they signed an agreement to deepen their collaboration to work together on further initiatives and analyses of towards being a sustainable brand.

The collaboration focuses on five main points which are further supporting San Benedetto to be one of the pioneers of a Circular Economy approach (De Giovanni, 2016). Firstly, taking continuous actions with the purpose of reducing and neutralizing carbon emissions in accordance with the Kyoto protocol and the Paris Agreement. Secondly, to promote innovations regarding Eco Design of their manufactured products and to exploit ecological efficient processes of their (management) processes (De Giovanni, 2020). Moreover, they also agreed on further analysis and optimization of the environmental production, distribution and marketing of their water. In this area they focus also on Scope 3 emissions and further indirect factors of the wider upstream and downstream supply chain. To tackle this issue, they wanted to include further associations to assess those raw material used and emissions more clearly. To further build out their knowledge base, San Benedetto also agreed on promoting joint information access to exploit also more external knowledge on how to reduce and assess the ecological footprint (De Giovanni and Vinzi, 2014). As a last point, they are elaborating together the environmental impacts of the production and selling of bottled water in general. One of these above-mentioned external associations is CSQA. This well-reputed company is responsible for validating and certificate the tracked progress regarding a reduced material use and steps toward being a Circular Economy (De Giovanni, 2019). All these points, especially the second, picture a big initiative and support for the circular economy approach at San Benedetto. With the support of the Government Side, in financial and reputational term, San Benedetto enjoys a top tier base. And it is no coincidence that it was San Benedetto who signed the agreement with the Ministry of Environment. Gian Luca Galetti, minister of environment, stated after singing the agreement that this agreement is only another sign of their appreciation on how San Benedetto focusses on being and becoming a sustainable company (Vishkaei et al., 2021).

How the Circular Economy Impacts the Corporate Strategy

Nowadays companies need new corporate strategies and models, in order to assure their profitability and competitive advantage over other producers. Moreover, preserving natural resources and reducing the negative impacts on the environment can be a useful and desirable weapon against the effects of climate change. These effects can contribute to the society's welfare and to people's well-being (De Giovanni and Zaccour, 2022). The circular economy is a unique opportunity for the firms to achieve these goals, plus to implement sustainability and the before-mentioned aspects into their corporate strategy. To execute these initiatives great commitment is badly needed to drive the necessary changes within the company's structure. Companies react differently to these new, external challenges based on their business models, business sector, and internal structure. The two possible reactions could be defensive or pro-active.

According to researchers adopting green practises has a significant impact on the financial and operational performance of the company. As the short-term effect, they can immediately reduce the costs, increase revenue and the efficiency of the processes through the reduction of waste and energy consumption, while they can become more appealing to customers and improve the overall perception of the company in the long run (Maranesi and De Giovanni, 2020).

The Links Between Secular Economy System and Performance

San Benedetto's corporate strategy has been built around the 4R principles (rationalization, reduction, responsibility, recycling). Rationalization means that San Benedetto is constantly investing into research and development, which has allowed them to adopt to new production processes and methods. The second point stands for reduction, which reflects San Benedetto's packaging system. The company reduces PET in the production of packaging resulting in lighter bottles, which is nonetheless an energy-saving method. Responsibility is about to decrease the CO2 emission of the firm. The company tries to finance green, eco-friendly projects to balance their CO2 offset. Through recycling it aims to give PET bottles new lives. Their range of products include packaging with RPET bottles.

The feedstock which leads us to the RPET is PET. As we know it, PET is a strong, durable, and recyclable material that is used to produce soda and water bottles. In order to produce RPET (which is not a virgin plastic) post-consumer PET has to be collected, sorted, and recycled. Then it must be refined into flakes that can be turned into new products, which is RPET in our case. Producing RPET requires 50% less energy than making PET from scratch and by using existing bottles we can be sure that these won't end up in a landfill. As we have already mentioned earlier, this enabled San Benedetto to produce significantly lighter bottles. The different kind of bottles (0,33l; 0,5l; 1,5l) were 37%, 41% and 21% lighter in 2017 compared the bottles from 2008.

San Benedetto's constant mission is to reduce the environmental impact and to offset green-house gas emissions (De Giovanni, 2020, 2022). In order to reach this goal, they introduced the Ecogreen line in 2012, made up with 50% plastic emission. The firm's commitment goes toward two main directions: on the one hand the reduction of carbon-dioxide emission, and on the other hand the recovery of PET.

After a change in the legislation, the company could reach a totally carbon-neutral production. Whereas previously they could only produce up to 50% carbon-freely, now they are able to create an "Ecogreen Easy Bottle" as carbon-neutral and 100% RPET made bottle. In addition, they can save 9% of greenhouse gases along the entire lifecycle of the product and nowadays San Benedetto produces 300 tons less of virgin PET.

The main goal is to reach a target set by the Europe Union (both in terms of recycling and minimum use of RPET) at 25% in 2025 and at 30% in 2029. Also, as a part of the Ecogreen initiative San Benedetto introduced twist and drink caps which are tied to the bottle and are not able to disperse into the environment, while at the same time being smaller, they contain less plastic than before.

San Benedetto anticipated the European Directive 2019/904, which requires the cap being attached to the bottle to be mandatory starting from 2024. This and the before-mentioned actions and innovative methods contributed to the fact, that the company could decrease its carbon footprint by 8826 tons between 2013 and 2020.

Furthermore, San Benedetto's environmental policy includes the choice of diversifying the production in 5 different, strategic location all around Italy (Scorzé, Popoli, Donato, Atella, Viggianello), which brings the production closer to the consumers. This is called "Project Network", a network which was designed to enhance the high-quality local water through Italy and helped to satisfy the consumers higher expectations connected to the supply of production (Duhaylongsod and De Giovanni, 2018). Not only gives this a flexibility to the company, but also helps to reduce the carbon emission of the company, as they do not have to transfer the products that much. The Project Network reduced the incidences on the roads and as the trucks had to run by 35,876km less which caused a 23,221 tons of CO2 reduction.

Circular Economy Developments and Business Transformation

From the very beginning, Acqua Minerale San Benedetto has always been trying to seek innovation processes, transforming the business on a regular basis. At the creation of San Benedetto in the 1950s, we can state that the supply chain of the company was, in a way, circular. On those days, the water extracted from Scorzè, Venezia, was distributed into glass bottles. After consumption, the bottles would head back into the supply chain to be transformed back again in bottles, ready to be reused (Jalali et al., 2020). However, using glass to transport water, comes with a lot of different issues. For instance, in winter, when reaching freezing temperatures, the glass will crack during transport or storage. Whereas during summertime, the pressure would blow the caps. In overall, shipping was both expensive and exhausting also because of the weight of the bottles that were packed into wooden crates. Then the first step to reduce cost and weight bottle production would be to use plastic crates instead of wooden. Another packing option for the glass bottles was using a wrapping plastic sheet. The first development in terms of business transformation of the packaging has been the reduction of the thickness of the wrapping plastic sheet going from 15 to 6 microns, allowing to a reduction of raw materials then costs.

The biggest transformation for the company has been to find a way to replace the glass bottles to simplify the supply chain and stop relying on powerful glass producers. The first attempt of this maneuver was made with PVC. But this type of plastic was not adapted to food and beverages purposes. Acqua Minerale San Benedetto had to find another type of plastic. The solution fits in three letters: PET for polyethylene terephthalate. The company decided to integrate the production of plastic bottle into the general process of production. Being able to control the creation from the PET granules to plastic tubes and finally blow air into the tubes that will form preselected molds. The transformation allows the company to create a bottle that is not subject to climate conditions and is, as a consequence, way lighter than glass bottles. Leading to a decrease of distribution costs due to a saving of fuel. This innovation also allows San Benedetto to integrate a great diversification of products. The company created new types and sizes and bottle to fit with their customers' needs. From the 0.5 L sports bottle to the 8 L bottle.

Since then, Acqua Minerale San Benedetto has implemented a proactive strategy called Ecolosophy. This philosophy aims to fulfill three main objectives: 1) Proximity to customers 2) Production efficiency 3) Diversification.

As mentioned above, starting from 1995, the company engaged an expansion strategy acquiring a new water production plant in Italy. Two years after, the expansion started to be international while investing in Spain then continued in 2001 with the acquisition of two other plants in Poland and Hungary. In 2021, the company owns 11 establishments in total to expand its market presence in Europe but also allows it to be closer to their direct customers. By owning these plants, the company tries to reduce as much as possible the number of kilometers the water must do to reach the customers. Leading to a reduction of CO_2 emissions. This decrease is significative because distribution represented more than a quarter of the Carbon Footprint of the company in 2016.

Another concern of this Ecolosophy strategy is the production efficiency. While addressing this topic, we should consider two main items: the materials and the energy used to produce a bottle. In terms of materials, for 10 years, San Benedetto has tried to reduce the quantity of plastic and the quantity of raw material needed for the production of a bottle. Leading to an average 20 percent reduction in the weight of the containers and up to 50 percent of recycled PET in all production countries as stated above. However, the biggest business transformation has been to offer to customers a bottle produced with 100 percent of recycled plastic only in Italy. Allowing a great improvement in terms of circular economy development. The next goal will be to create the same offer in other markets such as Spain, Poland and Hungary.

Offer diversification is also an important pillar of the business transformation because apart from plastic, San Benedetto is also offering other types of containers and distribution methods. One hundred percent recycled aluminum cans allow the company to offer a new type of product that could change the way of drinking water creating an endless and easy to recycle material. The Italian firm has also chosen to keep offering a premium glass bottle segment which is addressed to professionals such as restaurants or bars, but it is also offered to households. To control the entire supply chain, the glass bottles are returnable to the producer. Allowing San Benedetto to recycle a huge part of their bottles while reducing their GHG emissions to 9,9% from the glass bottles production (2013 vs. 2016).

However, globally speaking the company is still generating CO2 emissions that could be reduced. The next objectives for San Benedetto focus on the reduction of emissions related to distribution by limiting the use of thermic engines working with regular fuel. The use of methanol as fuel could be a new solution. It is important to heed note that controlling the full distribution chain is not an easy task because the last kilometer of the bottle; the one that pollutes the most; is not taken care by San Benedetto but by the final distributor or customer.

Finally, another improvement of the circular economy aspect could be to include in this meaningful strategy their most valuable partners such as Schweppes International and Pepsi-Cola. Especially when noticing that Coca-Cola has recently been criticized because of its large quantity of plastic used for its bottles. It is, moreover, crucial to encourage the end user whether to recycle as much as possible their containers or not using single-use containers. San Benedetto could study its next business transformation in this way by creating offers related to smart cities.

POSSIBLE IMPLICATIONS IN TERMS OF SMART CITIES

Smart cities adopt smart systems to improve the quality of life and to enhance sustainable development by making daily activities more easily and efficiently. For San Benedetto, a collaboration with the respective city regarding the following aspects is conceivable (Sacco and De Giovanni, 2019).

To know exactly how much plastic waste is generated by bottles, the volume of waste should be measured, as well as how much of it is being recycled. This could be implemented with sensors-based waste bins, which transmit the waste level status to the responsible bodies. Furthermore, it is important to have enough collection points everywhere in the city, especially where a huge amount of waste is generated by plastic bottles, i.e. in tourist places, in front of schools or universities. The collection points must be clearly recognizable so that sorting and the separation of the bottles appears easily accessible and appealing. To keep the cycle going, the collection points must be picked up regularly. A smart solution for the collection of the waste bins thus includes an optimized route of the collector. With sensors-based bins, unnecessary visits can be avoided and the best, most efficient and fuel-saving route is recomputed each time. (Aazam et al. 2016) The "New Life for PET" project already implemented some of these measures, including setting up various collection containers and collaborating with the plastic recycling company Aliplast S.p.A. to optimize the collection and recycling of the bottles. However, in order to move towards a circular economy, the implementation of smarter measures is needed. Especially in big cities like Rome, which currently struggles to handle solid waste, there can be indicated great potential for a more sustainable and efficient waste collection.

CONCLUSION

San Benedetto can be described as a resilient and reliable company, which through its history has already taken a pioneering role in plastic packaging in Italy. With regard of the long run aspects of the quest of San Benedetto, there are several proves that the company has steadily been increasing its reputation in the recent years. They act as a role-model for other non-alcoholic beverage companies, which culminated in the fact that in 2019 the Global Reputation Study selected them with the highest reputation index in Italy among mineral water companies. The company constantly invests in new innovations related to sustainability, as eco-sustainability is seen as a business mission. This is demonstrated by the voluntary agreement with the Italian Ministry for the Environment, the reduction of emissions through the "Network Project" and the "New Life for PET" project, which increased the recycling rate and encourages consumers to collect waste. Especially the agreement with the Ministry leaded to a deep collaboration. The Italian Ministry of the Environment appointed San Benedetto to help create the principles of Product Environmental Footprint of the sector and provide the criteria to evaluate the environmental performance. San Benedetto not only takes continuous action to reduce and neutralize carbon emission, but also invests in new manufacturing and continuously tries to optimize production, distribution and marketing. San Benedetto's knowledge and technological know-how have been consistently benefiting the environment, the economy and the people. Their implementation of Ecogreen line along with other eco-friendly initiatives are crucial in order to provide not only short-term results (cost reduction, product development, etc.), but also relevant long-term advantages as well, such as the image of an environmentally conscious company. For the future, San Benedetto should come up with smart solutions for waste collection, the strong cooperation with the ministry can help with the implementation in cities.

REFERENCES

Aazam, M., St-Hilaire, M., Lung, C. H., & Lambadaris, I. (2016, October). Cloud-based smart waste management for smart cities. In *2016 IEEE 21st International Workshop on Computer Aided Modelling and Design of Communication Links and Networks (CAMAD)* (pp. 188-193). IEEE. 10.1109/CAMAD.2016.7790356

De Giovanni, P. (2016). Coordination in a distribution channel with decisions on the nature of incentives and share-dependency on pricing. *The Journal of the Operational Research Society*, *67*(8), 1034–1049. doi:10.1057/jors.2015.118

De Giovanni, P. (2019). When feature-based production capabilities challenge operations: Drivers, moderators, and performance. *International Journal of Operations & Production Management*, *40*(2), 221–242. doi:10.1108/IJOPM-04-2019-0309

De Giovanni, P. (2020). *Dynamic quality models and games in digital supply chains*. Springer Nature.

De Giovanni, P. (2022). Leveraging the circular economy with a closed-loop supply chain and a reverse omnichannel using blockchain technology and incentives. *International Journal of Operations & Production Management*, *42*(7), 959–994. doi:10.1108/IJOPM-07-2021-0445

De Giovanni, P., & Vinzi, V. E. (2014). The benefits of a monitoring strategy for firms subject to the Emissions Trading System. *Transportation Research Part D, Transport and Environment*, *33*, 220–233. doi:10.1016/j.trd.2014.06.008

De Giovanni, P., & Zaccour, G. (2022). A selective survey of game-theoretic models of closed-loop supply chains. *Annals of Operations Research*, *314*(1), 1–40. doi:10.100710479-021-04483-5

Duhaylongsod, J. B., & De Giovanni, P. (2018). The impact of innovation strategies on the relationship between supplier integration and operational performance. *International Journal of Physical Distribution & Logistics Management*, *49*(2), 156–177. doi:10.1108/IJPDLM-09-2017-0269

Jalali, H., Ansaripoor, A. H., & De Giovanni, P. (2020). Closed-loop supply chains with complementary products. *International Journal of Production Economics*, *229*, 107757. doi:10.1016/j.ijpe.2020.107757

Mango, R. (2020). *Sustainable practices in plastic waste management – case study of Acqua San Benedetto SpA*. Masarykova univerzita, Ekonomicko-správní fakulta; Masarykova univerzita; Masaryk University. Retrieved from https://theses.cz/id/7f782u/

Maranesi, C., & De Giovanni, P. (2020). Modern circular economy: Corporate strategy, supply chain, and industrial symbiosis. *Sustainability*, *12*(22), 9383. doi:10.3390u12229383

PETnology. (2021). *Acqua Minerale San Benedetto launches the new Ecogreen 1l Easy Bottle, made with 100% recycled PET*. Retrieved from https://www.petnology.com/competence-online/news/acqua-minerale-san-benedetto-launches-the-new-ecogreen-1l-easy-bottle-made-with-100-recycled-pet.html

Sacco, A., & De Giovanni, P. (2019). Channel coordination with a manufacturer controlling the price and the effect of competition. *Journal of Business Research*, *96*, 97–114. doi:10.1016/j.jbusres.2018.09.001

Vishkaei, B. M., Fathi, M., Khakifirooz, M., & De Giovanni, P. (2021). Bi-objective optimization for customers' satisfaction improvement in a Public Bicycle Sharing System. *Computers & Industrial Engineering*, *161*, 107587. doi:10.1016/j.cie.2021.107587

Chapter 7
The Relevance of CIER in the Italian Circular Plastics District

Francesco Cuonzo
ENEL Group S.p.A., Italy

Melania Cocomero
KPMG, Italy

Paolo Carrassi
Luiss University, Italy

Vincenzo Casillo
Luiss University, Italy

Giulio Magnini
Berluti, Italy

Marco Madonna
Luiss University, Italy

Mariavittoria Spinosa
Luiss University, Italy

Giorgia Grassi
Luiss University, Italy

EXECUTIVE SUMMARY

The investment process in a circular economy project develops along a path that starts from the identification of the target project in which the preliminary objective and the investment opportunities are defined, consistent with the financial capacity of the company and with the role assigned to the project as part of the company's core business. Secondly, it is necessary to identify the strategic profile that the investment in this project carries out for the activities already progressively carried out by the company: since the objectives are the optimization of the profitability and liquidity of the investment, together with the returns of reputation competence. Therefore, the company's top management must be able to identify growth potential guaranteed by a solid business idea, also guaranteeing the sustainability of the flow of operations. The CIER company, now established in its sector, recycling of plastic materials, fits into this framework. This company is an exemplary example of how it is possible to develop and integrate a business project that is circular.

DOI: 10.4018/978-1-6684-5001-7.ch007

INTRODUCTION

This paper aims to analyze the circular district of CIER, a company that deals with plastic recycling. The recycling process by CIER starts with the purchase, through periodic auctions, of waste from two main national consortia (COREPLA and CONAI) that deal with the collection of plastic waste. Then, through a specific recycling process, CIER obtains a second high quality raw material, comparable to the characteristics of the raw material, which is sold to certain customer segments such as producers of food packaging or producers of classic packaging. This paper also analyzes the possible implications that a company like CIER has in the field of smart city and ecological transition.

CIRCULAR ECONOMY NETWORK

The circular process begins with the individual collection processes of urban and industrial recycling. COREPLA collects waste from associated landfills and organizes monthly auctions in which the companies of the category participate. CIER buys post-consumer plastic which is then selected by type of composition and color. The result of the process developed by CIER, through specific transformation processes, can be a scale, a granule or more frequently a coil. These products are then sold to partner companies that use the secondary raw material for their products.

The most important suppliers of CIER are:

- **Corepla**, National Consortium, a non-profit organization for the collection, recycling and recovery of plastic packaging, which brings together companies in the packaging supply chain. Despite being a private consortium, its purpose is of public interest: the achievement of the objectives of recycling and recovery of plastic packaging provided for by European legislation, with a view to shared responsibility between companies, Public Administration and Citizens.
- **CONAI,** "Consorzio Nazionale Imballaggi", a private Consortium that operates on a non-profit basis, which constitutes the response of private companies to a problem of collective interest, such as the environment, in compliance with guidelines and objectives set by the political system. About 760,000 companies producing and using packaging are members of the Consortium System. CONAI directs the

activity and guarantees the recovery results of 7 Consortia of materials: steel (Ricrea), aluminum (Cial), paper / cardboard (Comieco), wood (Rilegno), plastic (Corepla), bioplastic (Biorepack), glass (Coreve), ensuring the necessary connection between these and the Public Administration.

These bodies support the Municipalities in activating and developing adequate separate collection systems, recognizing them or the operators delegated by them the fees provided for by the ANCI-CONAI Framework Agreement to support the increased costs incurred for the performance of the collection. It also communicates to citizens and various stakeholders, raising awareness of the best practice of separate collection and, more generally, promotes interventions to reduce the environmental impact of plastic packaging starting from the companies belonging to the consortium.

The companies participating in the consortium are divided into 4 categories:

- **Category A** "Producers": mandatory adhesion by companies producing or importing raw materials for the manufacture of plastic packaging;
- **Category B** "Transformers": compulsory membership of companies producing or importing plastic packaging;
- **Category C** "Self-producers": optional membership by user companies that manufacture their plastic packaging or import packaged goods;
- **Category D** "Recyclers and recuperators": optional membership of companies that recycle and/or recover plastic packaging waste.
- waste prevention.

The different products made by Cier are destinated to three main different sectors:

- Packaging for cosmetics.
- Packaging for foods, such as bottles for water or packs for pasta.
- Other uses that are not destinated to foods packaging, such us the production of tubes, vases and other different objects.

They have also other minor customers which vary continuously and with whom are undertaken short term relationships, and also in this case they comes from different sectors.

Analysis of the Value, Drivers, Stakeholders, and Barriers

Corporate social responsibility (Porter's CSR) has its roots in the philosophy of Edward Fridman, who introduces "the theory of stakeholders" whether they are internal (shareholders, managers) or external as suppliers or customers (Sacco and De Giovanni, 2019). In this sense, the vision of interests that converge around the company is broadened. the company produces benefits for all interlocutors of the company, a common sense of value created by the company for the community emerges. The concept of "Corporate Good Citizen" is born, for which extra-business relationships are emphasized. The principle of good citizenship creates added value for stakeholders and is not based on ethical principles but on the desire to create public well-being through actions aimed at meeting economic, social and environmental objectives (triple bottle Line). The changes and innovations that the EC literature considers essential for the development of a sustainable Business Model (BMs), can be traced back to the studies of Vermeulen, who identified a series of actions that individual companies should pursue to adapt and integrate their Business Model with the current macro-trends of the global market (Jalali et al., 2020).

A sustainable corporate vision must include objectives such as the reduction of dependence on virgin raw materials and therefore the shift towards forms of supply from renewable energy, therefore the implementation of sustainable production practices and finally the adaptation of value chain strategies. CIER, the company dedicated to the recovery of plastic waste, represents an example of a sustainable corporate vision. This circular company was born in the two thousand years, with the aim of recycling plastic, expanding the range of action of the "DI GIACINTO" group, already active since the seventies in the recycling of metals. Today the twenty-year entrepreneurial reality of the family consists of seventy employees and a turnover of about thirty million euros per year. Framed the topic that we will deal with in this paragraph, listing the changes implemented in their own size and in the business model necessary to increase circularity, constituting a BMs.

Products that act as inputs to the company are fully recycled and reused, reflecting a balance of economic, ecological and social needs. CIER in the recycling of plastic materials performs its task, buying post-consumer plastic, that is the one already used that requires the recycling process. The finished product of the company is the second raw material or that obtained from the transformation process. The plastic is purchased according to the auction process by the two consortia that manage this system in Italy or COREPLA and CONAI. Offers to customers are sent through periodic auctions depending on the type of plastic to be purchased (PET, HDPE and LDPE).

The company's vision aims to create value by pursuing a strategy of "circular economy for real", the company's business begins in fact, with the recycling of plastic waste and ends with products first seconds. The production process produces material waste only in case of malfunctions or errors of the machinery, so there is no waste of material and related environmental impacts. The concept of "upcycling" comes to life, through a technology that through chemical treatments allows the reuse of materials with high added value, and represents an alternative solution to the disposal of waste by incineration. Compared to 41,000 tons of waste input, CIER has placed on the market in 2021 about 36,500 tons of output.

Resources and Capabilities

Products must be made through specific processes, with the help of innovative mechanisms. In addition, it is necessary to create reverse logistics systems which implies a renegotiation of the conditions with the carrier and the definition of closer partnership agreements, in line with the "end-of-waste" regulations designed and applied to promote the "Industrial Symbiosis" which provides for a collaborative and continuous interaction between companies, through the strategic exchange of inputs and outputs (Maranesi and De Giovanni, 2020).

In this sense, the outputs of the CIER company, (rPET, rHDPE and rLDPE), are transformed into stocks, coils or fragments, divided by color. The characteristics of the coils make the recycling process more or less complicated. The "light" coils are those with the greatest potential for recycling, while the darker, colored and opaque ones are those with less potential for recycling. Once the color distinction process is completed, bobbines are inserted into the recycling plant, while caps, shoppers and other plastic residues of lesser value are sent to the production of RDF.

The complexity of the supply chain is dictated by the difficulties that derive from the treatment of these materials, plastic is one of the most common materials, but also represents one of the most polluting and difficult products to dispose of among solid waste, since plastics in themselves are not biodegradable and, if incinerated, generate extremely toxic substances, such as dioxin. The ability of the CIER company to carry out activities aimed at creating shared value for companies and communities constitutes an added value that makes their output exclusive, in line with the principles of the RBW.

Activities and Drivers

CIER stipulates sales contracts in some sectors to ensure the required volumes of finished product to end customers. We note an increasing need of large international groups (preform producers in the food grade pellet user sector) to ensure feedstock in the coming years. This need arises from the progressive tendency to reduce the availability of recycled polymers, think of the virtuous "plastic free" habits.

The plastic waste once purchased at auction, at a variable price, for which the price of the finished output will also be variable, arrives loose or packed in the company and conveyed inside a huge shed, where the real selection takes place. The waste travels on belts at a decidedly high speed (3 m / s) and large puffs of air make a first selection, separating the heaviest from the lightest. Subsequently, the workers proceed to the manual sorting of that 3% that the machines cannot automatically separate. Once collected, the plastic is pressed to facilitate its transport to the plants that deal with the selection and recycling of plastic for its reuse as a "secondary raw material", that is, as a non-virgin raw material. From the collection of plastic, it is destined either to sorting and recycling centers or to compaction centers (Duhaylongsod and De Giovanni, 2018).

The plastic, once it reaches the recycling plant, is placed in a special system that provides for the tearing of any collection bags. The first selection of mechanical plastic takes place in a rotating screen that separates the various plastics according to their size. Following this first mechanical separation that is carried out by the rotating screen (a sort of huge centrifuge), further separations take place through optical readers that separate the plastic according to the polymers of composition and any colors. In this case the separation does not take place with a rotary movement (centrifugal) but with air puffs. At this point a second compression takes place with storage where homogeneous packaging is formed given by the same types of plastics. The output produced has different sales prices in relation to the competitive scenario. CIER is a company that aims to have an Ebit-Da > = 10 and therefore as a company with a good market performance. The current trends of scarcity of raw materials (the so-called "plastic free" determines less availability of plastic waste and consequently less material to be recycled) make MP very expensive and sales prices have been constantly increasing throughout 2021.

Outputs and Guarantees

CIER, through multiple machines and different process layouts, is able to treat seven types of plastics. Each is identified by a code that distinguishes its properties. We distinguish in this way: PET, HDPE, PVC, LDPE, PP (polypropylene), PS (polystyrene) and others. According to the characteristics of each, we distinguish different processes and consequently different outputs:

- **R-HDPE (HDPE)** from post-consumer bottles, mainly used in the production of membranes, tubes and articles for street furniture.
- **R-LDPE (LDPE)** from post-consumer film intended for molding agricultural articles, production of insulating sheaths and pipes.
- **R-PET (PET)** from post-consumer liquid containers and is used for the production of fibers, plates and packaging. Available in transparent, bluish, floral colors.
- **R-PET FOOD and R-PET Granule** are recycled products suitable for the production of packaging for food, liquids and solids, guaranteeing a degree of cleanliness identical to virgin as recognized by the "expert opinion" of Fraunhofer.

CIER is the first company in Italy to have completed EFSA's process for the production of R-PET suitable for contact with food and to receive the *EXA certification* or "European body that deals with certifying the dangerousness of plastic if in contact with food". The CIER company is aware of being a fundamental link in the customer consumption chain, therefore it operates following the "Lean Six Sigma" methodology, constantly improving processes to fully meet the customer's needs in quality standards while respecting competitiveness. The quality of the processes, the respect and protection of the environment and the safety of working spaces are guaranteed by the ISO9001:2008, ISO14001:2004, BS OHSAS18001:2007 and EMAS certifications.

Stakeholder, Costumers and Costumer's Interfaces

Circular economy actions manifest themselves at different levels in which a wide audience of stakeholders are involved (De Giovanni, 2016). The supply chain changes in order to involve a range of suppliers in the "sustainable supply chain management" (for material cycles). The sale of "circular" products may require first of all changes in customer habits or, if this is not possible, even a change in the customer portfolio, if it is not possible to motivate customers to take responsibility for their consumption. In this perspective, the certifications and awards obtained, like the environmental certification and the health certification, are all linked to quality dimensions, which in the eyes of the consumer represents a demonstration of a certain competitive advantage achieved by the company. To better understand what we are referring to, we use the definition provided by Fussler and James (1996), for which pursuing social objectives is the process that guides the consumer to prefer purchases of green products at a higher price in order to pursue a better social utility, allowed by high levels of transparency.

Economic Evaluation Analysis of the BM

The use of indicators such as the "Life Cycle Assessment" favors a more complete assessment of the environmental impact, on human health, with reference to ecosystems and resources; this indicator is all-encompassing as it observes both macro level (cities, regions, states); meso level (focus on supply chains and sectors and dynamics of industrial symbiosis); micro level (organisations, enterprises and products). It is possible to implement the analysis range of the LCA indicator alongside other indicators developed by the scientific community at the micro level.

In developing the economic assessment of BM, an "end-of-life-recycling rates" analysis is optimal by observing the end-of-life recycling rates of a product; also useful could be the use of the "Circular Economy index" indicator which highlights the difference between the economic value of the recycled material compared to the economic value of the "new" material used to produce a new version of a given product with reference to a specific material. Finally, it is reasonable to report the indicator "Sustainable Circular Index" which contains a set of indicators aimed at measuring how sustainable and at the same time circular a company is and the indicator "Material Stock Per Service" which instead measures the quantities of resources stored by a given company, necessary to make a given product (De Giovanni and Vinzi, 2014).

Buyers

Recycled plastic once produced in granule, flake or rPET coil is sold to two customer segments. Some use recycled plastic to create wrappers that are not brought into contact with food (about 40% output) for example manufacturers of jars, pipes, fruit boxes and even large segment of cosmetic packaging. Behind this choice of customers to buy recycled plastic there is a choice of sustainability since recycled plastic tends to cost more than virgin plastic since in addition to there being a high cost of processing and being there auction mechanism it can happen that the input cost is higher so it will also have to be higher cost of output. But there are also contracts stipulated for which the price is the same and therefore you have an average price. The other segment (i.e. 60%) is the one that produces packaging that comes into contact with food such as milk bottles, tortellini wrapper etc.

Recycled plastic is sold by CIER in many cases to companies that have the purpose of producing wrappers for food manufacturers or directly to multinationals that have presses to make containers.

The company's vision aims to create more and more a company attentive to the needs of the industry by offering products (recycled plastics) that contribute with their use to create a true circular economy by reducing CO2 emissions and giving the possibility of having finished products with an LCA (life cycle assessment) lower than the respective products only with virgin plastic polymers.

Barriers

There are several factors that hinder the implementation of circular economy business models, such as:

1. Technological skills;
2. Product portfolio;
3. Operational risks
4. Risks of cannibalization;
5. Financing the capex investment budget, as part of a strategic planning process, crucial for the performance of companies in terms of sustainability;
6. Social barriers.
7. Specialized staff and knowledge of the materials to be used;
8. European regulatory constraints.
9. Different types of customers with different needs

The environmental barriers, in terms of maximum emissions or pollution of the process, pose stringent constraints, which do not represent a limit to the activity of CIER, but a due attention to the fact of having to do things "well".

Among the economic barriers, i.e., financial barriers and in terms of investment, for a company destined for the recycling of "end of life" products at this time it is basically easy to find financing for the initial phase: the plastics recycling sector is seen as a sector in enormous growth both by the banking system and by European public funds (NRP).

Social barriers, related to the impact on employees and the community in case of non-compliance with production standards, represent a great danger for the company due to the risk of leakage of dioxin and other toxic substances. Another difficulty is represented by the trend to PLASTIC FREE of the current market, which influences the choices of packaging materials by large-scale distribution (retail) and industry (De Giovanni, 2020).

Analyzing the cost cluster, we estimate how for CIER these amount to about 10 million euros: 3.5 mil € personal company costs, about 3.8 mil € of energy and another 70% of turnover as MP. CiER's turnover in 2021 will be €44.5 million; capital needed to meet the costs of maintenance and supply chain implementations.

The studies of Guldmann and Huulgaard identify the main obstacles to the barriers at the Socio-technical level (organizational, value chain, level of employees, market and institutions).

How the Circular Economy Impacts the Corporate Strategy

Usually, the investment process in a circular economy project is developed according to a path that starts from the identification of the target project in which the preliminary objective and investment opportunities are defined, which is consistent with the financial capacity of the company and with the role attributed to the project within the company's core business. Secondly, it is necessary to identify the strategic profile that the investment in this project carries out for the activities already progressively carried out by the company: since the objectives are the optimization of the profitability and liquidity of the investment, together with the returns of reputation competence. Therefore, the most attractive projects are those with a management capable of identifying growth potential guaranteed by a solid business idea, also ensuring the sustainability of the deal flow.

The "Di Giacinto" group also in this case represents an exemplary case, since the CIER company was born as the last branch of a group, since the 70s, active in the field of metal recycling through the Metalferro company. Metalferro, is the parent company of other industrial companies specialized in the various sectors of collection and processing of industrial waste: C.I.E.R., Lucania Metalli srl and S.A.R.R.ME.

Thanks to the use of cutting-edge technologies and infrastructures, most of the processes are now fully automated and place Metalferro at the top of the national for production capacity and quality. The economic and financial results speak of an industrial force that grinds a turnover of over 50 million a year and has about 350 employees. From our research it emerges that the backbone of this success were the monitoring activities carried out.

The checkup, easily implemented with the help of the blockchain, has a pivotal function, as first of all it has a control role which is flanked by a supporting role of the project favoring the growth of skills relationships with intermediate and final customers, suppliers, partners, stakeholders, allowing to guarantee a favorable outcome (way out) by achieving the set objectives.

In this perspective, it is therefore necessary to identify fundamental drivers that define how a circular investment impacts on the business strategy. First of all, it is necessary to evaluate the relationship between an investment and the company's core business, since synergies can be identified with the financing company which guarantee greater possibilities of internalization of the start-up.

Subsequently, it is necessary to identify the possibilities that this new business has to be independent of the activity of the parent company, also evaluating the cost of skills, resources and specific talents so that the start-up can achieve this autonomy. In this sense, organizational contexts and corporate culture heavily influence the company's ability to transform different types of innovation into business, building groups (skunk works) isolated from the rest of the company that works on the development of innovation.

In this context, the degree of autonomy of the research and development of these prototypes must however be framed in a system that can allow a direct control both in terms of progress and governance, controlling the monthly costs, reporting, protocols but also the organization of the project team and the security of the latter in terms of confidentiality until the moment of internal proof of the real potential of the projects developed. Note how, in these terms, not only the business model and the internal organizational chart of the company and the network of suppliers and stakeholders vary and expand, but also the organizational environment, the level of risk, technologies and patents in the weighting of go-to-market alternatives.

THE LINK BETWEEN CIRCULAR ECONOMY AND FINANCIAL PERFORMANCE

A circular economy system can be linked to performance in three main aspects: financial aspect, environmental aspect and social aspect. This paragraph shows the possible links between financial performance and circularity. The Ellen MacArthur Foundation (MacArthur, 2013) describes four ways that companies which implement a circular business model can create value and compares those with companies that use a linear business model. The first possibility to create value is "the power of the inner circle" and compares circular material usage to linear ones. The second one is called "the power of circling longer", and it can create value by maximizing the number of consecutive cycles of products. Thirdly, the possibility to create value is called "the power of cascaded use", that includes the chance to reuse the materials into different supply chains (De Giovanni and Folgiero, 2022). Finally, the fourth possibility of value creation is mentioned as "the power of pure circles", in fact "Purer material streams more accessible to collect, redistribute and recycle, and result in a higher efficiency in the circular supply chain, extends product longevity and increases material productivity".

The financial performance of a company can be affected positively because it gains the capability to reuse and recycle resources. The main benefit is a cost saving and a reduction of environmental impact. There are some main drivers that explain why a circular business model can create a positive financial effect. The first driver is the possibility to gain additional revenue streams. This can happen as a consequence of different factors such as the creation of a long-term relationship between the customer and the manufacturer, the opportunity to make profit by selling the waste to other companies that can use it as a resource, and last but not least the growing importance that customers are giving to green products, due to which they prefer to buy environmentally friendly products1. The second driver regards the chance to reduce supply risk and price volatility through a well-developed internal resource management. The positive environmental attitude of circular companies is also useful to mitigate their reputation and regulation risks.

Table 1. CIER's financial statement, 2018

	31/12/2018	31/12/2017	Variazione
Ricavi netti	21.668.827	20.974.788	694.039
Costi esterni	19.143.762	18.512.032	631.730
Valore Aggiunto	**2.525.065**	**2.462.756**	**62.309**
Costo del lavoro	1.684.997	1.599.392	85.605
Margine Operativo Lordo	**840.068**	**863.364**	**(23.296)**
Ammortamenti, svalutazioni ed altri accantonamenti	790.432	987.520	(197.088)
Risultato Operativo	**49.636**	**(124.156)**	**173.792**
Proventi non caratteristici	128.330	600.202	(471.872)
Proventi e oneri finanziari	(266.989)	(312.743)	45.754
Risultato Ordinario	**(89.023)**	**163.303**	**(252.326)**
Rivalutazioni e svalutazioni			
Risultato prima delle imposte	**(89.023)**	**163.303**	**(252.326)**
Imposte sul reddito	42.483	80.961	(38.478)
Risultato netto	**(131.506)**	**82.342**	**(213.848)**

Finally, the third driver regards a social aspect: the promotion of sustainable practices by the company has a positive impact on employees, especially on their organizational commitment and their behavior as citizens.

CIER's Financial Performance

Analyzing CIER's financial statement, it is possible to notice how the company has grown during the recent years. Considering years 2017, 2018, 2019 and 2020, the main profitability indices appear to have generally grown at a stable pace. In 2018 the company incurred in higher costs due to investments to open a new production plant. In fact, it is possible to notice how in 2018, when compared to 2017, the company has recorded an overall loss at the end of the year.

Table 1 reports the comparison between 2018 and 2017 reclassified income statement.

Also, the profitability indices have reduced in 2018 when compared to 2017, as resulting from Table 2.

The same trend can be recorded when it comes to solidity indices, such as fixed asset to equity capital ratio (quoziente primario di struttura) and fixed asset to equity capital and medium-long term debt ratio (quoziente secondario di struttura), as shown in Table 3.

Table 2. CIER's financial indicators, 2018

	31/12/2018	31/12/2017
ROE netto		0,03
ROE lordo		0,06
ROI	0,01	0,02
ROS	0,01	0,02

Table 3. CIER's Financial margins, 2018

	31/12/2018	31/12/2017
Margine primario di struttura	(4.435.741)	(4.302.854)
Quoziente primario di struttura	0,37	0,39
Margine secondario di struttura	(2.178.404)	(1.113.721)
Quoziente secondario di struttura	0,69	0,84

Analyzing now years 2019 and 2020, the overall trend is a stable growth. It is possible to notice how CIER has increased its sales volume in the years and also its final profit, as resulting from Table 4.

As concerns the profitability indices, the results are displayed in Table 5.

Table 4. CIER's financial statement, 2020

	31/12/2020	31/12/2019	Variazione
Ricavi netti	28.430.320	25.309.924	3.120.396
Costi esterni	23.877.251	21.581.486	2.295.765
Valore Aggiunto	**4.553.069**	**3.728.438**	**824.631**
Costo del lavoro	2.921.588	2.399.127	522.461
Margine Operativo Lordo	**1.631.481**	**1.329.311**	**302.170**
Ammortamenti, svalutazioni ed altri accantonamenti	1.329.881	776.245	553.636
Risultato Operativo	**301.600**	**553.066**	**(251.466)**
Proventi non caratteristici	510.833	146.906	363.927
Proventi e oneri finanziari	(278.529)	(343.971)	65.442
Risultato Ordinario	**533.904**	**356.001**	**177.903**
Rivalutazioni e svalutazioni			
Risultato prima delle imposte	**533.904**	**356.001**	**177.903**
Imposte sul reddito	106.609	15.283	91.326
Risultato netto	**427.295**	**340.718**	**86.577**

Table 5. CIER's financial indicators, 2020

	31/12/2020	31/12/2019
ROE netto	0,07	0,07
ROE lordo	0,09	0,07
ROI	0,03	0,03
ROS	0,03	0,03

The solidity indices demonstrate how the company has become more stable in financial terms over the years, getting to a good solidity structure overall, as reported in Table 6.

Sustainable Development Goals (SDGs)

analyzed the 17 Sustainable Development Goals (SDGs), the main one that CIER is pursuing is Goal N° 12, shown in Figure1, "Ensure sustainable consumption and product patterns".

As it is shown from the image, "material footprint" is a massive problem, and we can notice that between 2000 and 2017 the global consumption of plastic increased enormously and most of it is thrown away, causing more pollution. That's why CIER's activity in recycling plastic is fundamental for the environment.

Some targets interesting to examine are:

- Target 12.1: "Implement the 10-year framework of programmes on sustainable consumption and production, all countries taking action, with developed countries taking the lead, taking into account the development and capabilities of developing countries". Through the following indicators: 12.1.1 (Number of countries developing, adopting

Table 6. CIER's financial statement, 2020

	31/12/2020	31/12/2019
Margine primario di struttura	(7.297.447)	(5.349.255)
Quoziente primario di struttura	0,46	0,51
Margine secondario di struttura	(1.208.916)	(3.845.415)
Quoziente secondario di struttura	0,91	0,65

Figure 1. SDG 12

or implementing policy instruments aimed at supporting the shift to sustainable consumption and production). This point is particularly consistent with current legislation that obliges plastic packaging manufacturers to include a minimum percentage of recycled plastic inside the packaging;

- Target 12.2: "By 2030, achieve the sustainable management and efficient use of natural resources". That is measured with the following indicators: 12.2.1 (Material footprint, material footprint per capita, and material footprint per GDP) and 12.2.2 (Domestic material

Table 7. CIER environmental declaration

CIER srl – Dichiarazione Ambientale

N°	OBIETTIVO	TARGET	DATA TERMINE	AZIONI PREVISTE E RISORSE	RESP.	PIANIFICAZIONE	STATO DI RAGGIUNGIMENTO
1	Riduzione Inquinamento atmosfera	Mantenere, per tutti i punti di emissione, il valore (medio e di picco) delle polveri<0,5mg/Nm^3 Val. attuale max: 0,8	31/12/2020	Manutenzione periodica del filtro di abbattimento e sostituzione pale nelle valvole previste. Costo previsto: €5.000,00	RSAL	**T_1-31/12/2018**: Valori polveri < 0,7 **T_2-31/12/2019**: Valori polveri < 0,6 **T_3-31/12/2020**: Valori polveri < 0,5	Il traguardo T_1 non è stato ancora raggiunto a causa di una manutenzione non completa.
2	Riduzione inquinanti scarico acque	Riduzione del **30%** di COD (V_{medio}) allo scarico (valore attuale 350)	30/06/2020	Definizione di un piano di manutenzione del depuratore a fine di migliorare il processo di filtraggio e pulizia delle acque.	RSAL	**T_1-30/06/2018**: riduzione del 10% **T_2-30/06/2019**: riduzione del 20% **T_3-30/06/2020**: riduzione del 30%	**T_1**: Raggiunto (-10%) **T_2**: Raggiunto (-20%) (Vmedio<30%)
3	Miglioramento consumi	Diminuzione del consumo di gas metano del 30% (valore attuale 502mila m^3)	30/06/2020	Ottimizzare l'utilizzo delle caldaie e dell'impianto di cogenerazione.	RSAL	**T_1-30/06/2018**: riduzione del 10% **T_2-30/06/2019**: riduzione del 20% **T_3-30/06/2020**: riduzione del 30%	**T_1**: Raggiunto **T_2**: Raggiunto
4	Riduzione inquinamento atmosfera	Abbattimento delle polveri/residui di lavorazione della Linea PE	31/12/2019	Installazione di un aspiratore sull'impianto Intarema. Costo previsto: €10.000,00	RSAL	**T_1-31/03/2018**: Individuazione dell'installatore. **T_2-31/12/2018**: Progettazione della linea di aspirazione. **T_3-31/12/2019**: Realizzazione della Linea di aspirazione.	**T_1**: Raggiunto **T_2**: Raggiunto

REV: 10 del 30 Novembre 2019 **Data 1°Emissione:** 24 Ottobre 2011 **Pagina 34 di 42** RINA

consumption, domestic material consumption per capita, and domestic material consumption per GDP);

- Target 12.5: "By 2030, substantially reduce waste generation through prevention, reduction, recycling and reuse". The indicators are: 12.5.1 (National recycling rate, tons of material recycled).

Cier's Environmental Declaration

Regarding the environmental declaration CIER srl follows the UNI EN ISO 9001, UNI EN ISO 14001, OHSAS 18001:2007 and EMAS VI standards. The company has identified some important environmental impact areas that are: raw materials, water resources, energy, ground, air, water, land, waste, emergencies. For each impact area they have provided some improvement measures (see Table 7).

In this chart there are some environmental goals, their targets, actions and resources needed to achieve the goal and the relative achievement status.

Table 8. Frequencies and gravity indexes

	2016	2017	2018	2019 al 30/09/19
Indice di Gravità (n°g. inabilità temp./N°h lavorate)*10³	0,29	0	0	0
Indice di frequenza (n°Infortuni/N°h lavorate)*10⁶	19,56	0	0	0

Tabella indicante gli indici di gravità e frequenza per CIER nel triennio di riferimento

Fonte dei dati: Registro Infortuni aziendale

Cier's Production

CIER srl is authorized to recycle as an input 62.000 tons of plastic per year. In 2021 the company has utilized about 41.000 tons of plastic waste as an input, so it has a chance to continue to grow as concerns the tons of plastic recycled. In 2021 with 41.000 tons of plastic waste as an input the company has produced an output of 36.500 tons circa of recycled plastic. The division of tons between the company's different outputs stands as follows:

- 5.300 tons of PET flakes
- 12.500 tons of Pellets food grade
- 9.700 tons rHDPE and rLDPE grain
- 9,000 tons Lamina PET, Pet-PE, Pet-Evoh-Pe

Social Aspects

CIER has decided to implement a health and safety management system respecting the OHSAS 18001:2007 standard. The company claims that the safety and health of its employees is fundamental to pursue a correct environmental policy. Table 8 shows the severity and frequency indices for CIER in the three-year reference period:

In its Ethical Code, following the legislation 231/2001, CIER affirms that it rejects all discrimination on the basis of age, gender, sexuality, health status, race, nationality, political opinions, and religious beliefs; repudiates all forms of discrimination in recruitment policies

and in the management of human resources. It is committed to preventing all forms of bullying and exploitation of labor.

For CIER it is fundamental to value human resources. At the moment, the company has a total of 72 employees with 49 permanent contracts. The personnel turnover is generally under 15% per year 2.

CIER: Circular Economy Development and Business Transformation

Faced with a market that is interpreting environmental sustainability as a real added value, the prospects linked to the circular economy are called upon to play a strategic role in many sectors: "doing good" for the environment is no longer just an ethical objective or linked to social responsibility but is increasingly a requirement for developing new forms of competitiveness and for meeting the needs of customers and markets.

The circular economy brings concrete benefits both to the environment and to business development as well as to the corporate profitability thus makes it possible to build a more resilient supply chain: in fact, it tends to be based on shorter, local and more flexible supply chains than global supply chains based on the principles of efficiency and with a higher supply risk, and it incentivizes locally mechanisms of repair and regeneration of the products without the need for supplies from other production chains (De Giovanni, 2022).

A more local supply chain is less affected by exogenous phenomena and reduces environmental impacts at least for transport activities; in this case, there is a more transparent supply chain which emphasizes the concept of visibility. Information must be visible in all points of the supply chain, and the integration of information of all the actors involved allows data to travel faster, to make decisions based on more precise knowledge: it allows to be more proactive and to introduce predictive concepts and, by virtue of this, also to reduce waste that is related to wrong decisions (De Giovanni, 2022).

As a partial demonstration of this, the circular economy is (and must) become a business opportunity for companies: the latest GreenItaly report highlighted how "green" companies have achieved better results (16% increased turnover vs 9% non-green); they innovate more (73% versus 46%); they invest more in R&D (33% vs 12%) and are more digitized. In fact, circular investments are associated with new sources of revenue, a reduction in costs and constitute a lever of differentiation from the competition, going beyond greenwashing for advertising purposes (Green Italy Report, 2021).

The development of the circular economy is related to economic growth: the greater efficiency in the use of resources, in extending the useful life of products and in reconditioning at the end of use, generate greater value as well as reduce costs. Some companies have started to ride these alternatives by generating new business models and opportunities: the company CIER is one of these.

At the moment, the company is specialized in the processing of low- and high-density polyethylene, coming from separate waste collection or from pre- and post-consumer industrial waste: it produces densified and ground granules of HDPE, LDPE, and Polypropylene after the recovery phase of the selected plastic material. Since March 2002, CIER has started an important economic and technological growth, investing considerable capital to renew and expand machinery and equipment, progressively increasing its turnover, in a historical period in where the value of plastics was in sharp decline.

From 2007, CIER started the production of PET (Polyethylene terephthalate) through 2 highly technological plants; the production of PET represented a strategic objective for the Management as the market is very large relative to the former. Furthermore, in 2010 CIER started producing PET for food-use by carrying out a further purification process through a highly technological plant. Thanks to the installation of the Bandera technological system, the production of coextruded rigid film in PET also began in 2015, used to cover trays and containers for food use. In 2016 a second production line joined the first.

In February 2019, the company also obtained certification to the BRC IoP global standard for packaging and packaging materials; this new certification will allow the company to align itself with the best production standards on the safety of food containers.

In the production cycle, CIER carries out processes of:

- Recovery and grinding of plastic materials;
- Production of PE and PET plastic granules and flakes also for food-use
- Production of PET rigid film.

The recovery and grinding are carried out in the plants in order to obtain the production of granules and flakes. In particular, in CIER there is:

Figure 2. Circular Economy at CIER

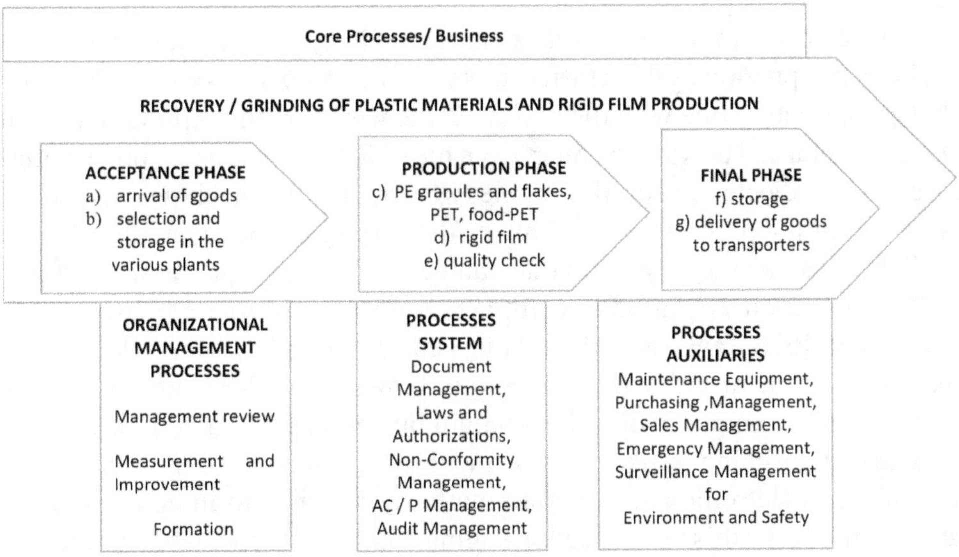

- a plant for the production of Polyethylene (PE) granules used mainly for the building accessories industry;
- two plants for the production of Polyethylene Terephthalate (PET) flakes used for various types of industrial applications with the exception of food or drink containers;
- a plant for the production of PET Food-Contact used for the production of containers in direct contact with food or drinks.
- a plant for the production of coextruded rigid film in PET.

All the previous production activities are carried out in compliance with the Provision of the Province of Teramo nr. 2037 of 17/10/13 authorizing the continuation of the non-hazardous waste recovery activity, pursuant to Legislative Decree 152/06 and subsequent amendments.

The business model and manufacturing process of CIER can be schematically represented as follows in Figure 2.

Starting from February 2021, CIER increasingly grew in terms of circular economy and its developments: it has equipped itself with a new machine for the optical sorting to maximize the yield of the plant: by doing this, it has expanded the recycling plant in Castelnuovo Vomano with a new line for the recovery of PET fractions: the company, which already has two sorting machines, recently added a new one, Mistral Plus, that will allow further recovery of those fractions that were discarded, ensuring complete recycling of PET waste.

The need was to optimize waste and completely recover the fractions of PET that had previously been lost: in this way, the company expanded the main PET pre-selection line with the new machine which removes pollutants such as trays, metal and unwanted colors from the PET flow. The new optical sorter receives the rejected material from the two Mistral DVI machines, which are upstream of the process, and optimizes the sorting and recover 100% of the PET that has been rejected and that can be used in other processing stages.

The sensors it is equipped with, are able to discriminate between PET bottles and PET trays, as well as being equipped with a sensor for metals that prevents the machine from selecting (blowing) bottles contaminated by metals or for example bottles that contain blister packs of medicines inside.

The new machine is also equipped with a central nervous system (CNS), which is an electronic and software platform designed to integrate sensors and technologies: this new selection engine reinforces the scalability concept, helping customers keep it constantly updated. The CNS expands the sorting applications and makes it possible to differentiate more complex materials, such as separating PET bottles and trays, or to distinguish paper from cardboard.

The company is growing at the level of 20% per year in terms of volumes and therefore is capable to recycle more and increase the size of the circular economy in Italy.

The business changed over the years: the legislative drivers that oblige companies which transform and produce packaging to use mandatory minimum percentage of recycled plastics, increased the size of CIER's final market.

The Italian and European retail trend that requires a guaranteed 60% to 80% of recycled plastic in the trays sold, also increases the use of recycled plastic produced by CIER; in short, a sharply growing market, constrained by a reduction in the availability of raw materials (waste to be recycled).

The company will face continuous transformations in the next few years in terms of development and business transformation: the sector faced a three-year period of strong growth with rising sales prices, in a scenario characterized by a very sharply rising of raw material costs due to the scarcity of raw materials. CIER will use less virgin plastic and will introduce other technologies such as chemical recycling to have more raw material to recycle (CIER's sector is now only in mechanical recycling). Recyclers will be incorporated into the supply chains of the large converters (packaging producers) who, by integrating, will ensure the feedstock of recycled material so essential to their business.

CIER: Possible Implications in terms of Smart Cities

It is now well known that one of the most pressing environmental challenges is reducing plastic waste globally: the continuing accumulation of plastic waste in landfills and oceans is threatening the planet's balance.

Research and companies together are addressing the issue and they are trying to find solutions to the problem of safeguarding and optimizing such a vital resource. The Mediterranean is one of the seas most affected by plastic pollution, reaching record levels of microplastics that threaten all marine life, including cetaceans, sea turtles and fish, as well as our health: for this reason, the WWF Mediterranean Marine Initiative (MMI) has launched a campaign in the Mediterranean region by putting pressure on governments, engaging with municipalities, collaborating with industry and mobilizing citizens so that by 2030 no more plastic is dumped into the sea (WWF Italia Onlus, 2021).

The "Plastic Smart Cities" project promoted by the WWF is perfectly positioned in this context and represents an excellent opportunity for CIER: in Italy, the municipality of Venice carried out this project to implement best practices that prevent, minimize, and manage plastic, both as a resource and as waste.

The strategic position of CIER and the use of innovative machinery for the recycling of PET material and the production of PET for food-use, would guarantee to the company a strategic role in the implementation of this project in collaboration with the municipalities of the Abruzzo coast.

In cities, will be discussed and developed innovative solutions concerning the plastic life cycle chain and developed: for example, the citizens could thus take their plastic waste to the 'waste banks' in exchange for a small reward, or many bins could be added to efficiently collect and recycle waste at coastal areas, ports, promenades, stations, and roads. Many restaurants and bars could offer meals and drinks in recyclable containers; citizens and tourists could enjoy nature and participate in significant cultural events with a plastic-free approach.

Cities and the WWF would engage in public awareness initiatives to inform and mobilize people to protect the sea from plastic pollution: the ambition of the project is the creation of a Mediterranean platform composed of the different municipalities of the Abruzzo coast in order to promote collaboration and the sharing of knowledge on how to implement effective and coherent solutions to avoid single-use plastics, obtain a complete collection system and recycling of plastics and reducing pressure from the tourism sector.

Hypothetically starting from the city of Pescara, WWF will collaborate with the critical sectors of the different cities (including industry, environment, waste management, circular economy/sustainability, tourism) to develop an analysis of the city in the field of plastics (from consumption to waste management) and therefore identify the primary sources of dispersion of plastic waste in nature. This approach will allow the city to develop an action plan with a series of concrete initiatives to reduce plastic pollution, adequately measure interventions and achieve zero-in-nature dispersion by 2030.

CONCLUSION

In conclusion CIER is a company of national importance for its role in transformation and recycling, in fact through its machinery, the company is able to provide its customers with a high-quality product comparable to virgin raw material. CIER's customers choosing to be part of this district undertake a sustainable choice, which requires a high and continuous investment because the price of secondary raw materials is higher than virgin ones. The company stands out on national soil for its significant contribution to the ecological transition, in a sector, that of plastics, which is in continuous growth. Through investments in innovative processes of Research and Development the company is fully integrated in the district, ensuring a process of Up-Cycling of plastic waste that causes high environmental impacts for their traditional disposal in incinerators.

In conclusion, CIER is a perfect case of a company that actively participates in the ecological transition, implementing a plastic recycling process that manages to create added value by creating zero waste.

REFERENCES

De Giovanni, P. (2016). Coordination in a distribution channel with decisions on the nature of incentives and share-dependency on pricing. *The Journal of the Operational Research Society*, *67*(8), 1034–1049. doi:10.1057/jors.2015.118

De Giovanni, P. (2019). When feature-based production capabilities challenge operations: Drivers, moderators, and performance. *International Journal of Operations & Production Management*, *40*(2), 221–242. doi:10.1108/IJOPM-04-2019-0309

De Giovanni, P. (2020). *Dynamic quality models and games in digital supply chains*. Springer Nature.

De Giovanni, P. (2022). Leveraging the circular economy with a closed-loop supply chain and a reverse omnichannel using blockchain technology and incentives. *International Journal of Operations & Production Management*, *42*(7), 959–994. doi:10.1108/IJOPM-07-2021-0445

De Giovanni, P., & Folgiero, P. (2022). *Strategies for the Circular Economy*. Circular Districts and Networks. Giappichelli-Routledge.

De Giovanni, P., & Vinzi, V. E. (2014). The benefits of a monitoring strategy for firms subject to the Emissions Trading System. *Transportation Research Part D, Transport and Environment*, *33*, 220–233. doi:10.1016/j.trd.2014.06.008

De Giovanni, P., & Zaccour, G. (2022). A selective survey of game-theoretic models of closed-loop supply chains. *Annals of Operations Research*, *314*(1), 1–40. doi:10.100710479-021-04483-5

Duhaylongsod, J. B., & De Giovanni, P. (2018). The impact of innovation strategies on the relationship between supplier integration and operational performance. *International Journal of Physical Distribution & Logistics Management*, *49*(2), 156–177. doi:10.1108/IJPDLM-09-2017-0269

Fussler, C., & James, P. (1996). *Driving eco-innovation: a breakthrough discipline for innovation and sustainability*. Financial Times/Prentice Hall.

Jalali, H., Ansaripoor, A. H., & De Giovanni, P. (2020). Closed-loop supply chains with complementary products. *International Journal of Production Economics*, *229*, 107757. doi:10.1016/j.ijpe.2020.107757

MacArthur, E. (2013). Towards the circular economy. *Journal of Industrial Ecology*, *2*(1), 23–44.

Maranesi, C., & De Giovanni, P. (2020). Modern circular economy: Corporate strategy, supply chain, and industrial symbiosis. *Sustainability*, *12*(22), 9383. doi:10.3390u12229383

Green Italy Report. (2021). *Un'economia a misura d'uomo per il futuro dell'Europa, Oct. 20th, 2021*. Author.

Sacco, A., & De Giovanni, P. (2019). Channel coordination with a manufacturer controlling the price and the effect of competition. *Journal of Business Research*, *96*, 97–114. doi:10.1016/j.jbusres.2018.09.001

Vishkaei, B. M., Fathi, M., Khakifirooz, M., & De Giovanni, P. (2021). Bi-objective optimization for customers' satisfaction improvement in a Public Bicycle Sharing System. *Computers & Industrial Engineering*, *161*, 107587. doi:10.1016/j.cie.2021.107587

WWF Italia Onlus. (2021). *Plastic Smart Cities*. Available at https://www.wwf.it/cosa-facciamo/progetti/plastic-smart-cities/

Chapter 8
Sustainable Fast Fashion:
Business Case of H&M

Giammarco Pastore
Intellera Consulting, Italy

Paolo Bernardo Di Mezza
Luiss University, Italy

Alessandro Frangiosa
Itabus, Italy

Maria Francesca Falcone
KPMG, Italy

Vincenzo Loria
Luiss University, Italy

Maria Romeo
Mercedes Financial Services, Italy

Matteo Bellomo
Urban Vision S.p.A., Italy

Lucie Marie De Cazotte
Luiss University, Italy

EXECUTIVE SUMMARY

Sustainability is significantly important to the fashion business because of consumers' growing awareness of the environment. When a fashion company aims to promote sustainability, the main link is to develop a sustainable circular system. This chapter contributes understanding of how a historical fashion company has managed to evolve over time by implementing circular technology that can give it a competitive advantage in the market. The authors firstly describe the structure of H&M, the value it distributes in the market, and the future goals it has set for itself. Next, they introduce the methodology by which the company has managed to make its circular model impact on its performance, giving an overview of the relationship the company has with its stakeholders and consumers. Furthermore, based on secondary data and analysis, they learn how the Swedish fast fashion company has built its sustainable strength by developing eco-friendly materials, monitoring sustainable production, reducing carbon emissions in distribution, and promoting circular fashion.

DOI: 10.4018/978-1-6684-5001-7.ch008

INTRODUCTION

As reported by Forbes, "fast fashion is the idea of moving large volumes of merchandise from the design table to the showroom shelves in the shortest possible time."

It's a business model that makes speed its strong point, according to a clothing supply chain model that aims to respond quickly to the latest fashion trends by frequently updating clothing products, with new styles being introduced frequently. Leading Fast Fashion retailers offer precisely low-cost collections that mimic current luxury fashion trends, satisfying customers' insatiable demand for novelty. In this way, companies have transformed themselves to meet the needs of modern consumption, which has evolved to an impulsive purchase of fashion. The production of Fast Fashion companies is essentially based on two logics, the pull logic, which allows to activate the production since the actual needs of the market and the Just in Time logic, which allows to minimize stocks, producing only what is expected to sell in a very short time or that has already been sold. The combination of these two logics guarantees the speed that is the defining element of this business model.

Among the leaders of fast fashion, a dominant position is undoubtedly held by the Swedish company H&M, acronym of Hannes & Mauritz, founded by Erling Persson in 1947. H&M has a distinct business model in that, it does not produce its products in-house but outsources its production to more than 900 independent suppliers around the world, mainly in Europe and Asia. H&M's critical factor, unlike other fast fashion players, is not the poor quality of its products and/or their durability, but rather the environmental and social impact of its production and distribution. It is precisely on these issues that the company's new business strategies are based, and which we will examine in detail in this work.

H&M's commitment in this direction, as we will see later, is considerable. In fact, the company is a member and sponsor of the Global Fashion Agenda to promote sustainable fashion, a member of the 60 brands that signed the Fashion Pact, with the aim of making fashion more environmentally sustainable, and it is also the creator and sponsor of the Global Change Awards, also created to reward sustainable ideas in the fashion industry.

The company, making sustainability one of its pillars, offers its customers the opportunity to participate in a circular process where all stakeholders are involved. Thanks to the conscious line, to the recycling of used clothes and to its sustainable choices, it turns out to be a good model to emulate in the industry and in society. In fact, implementing the H&M model within a smart city represents a very interesting perspective for the future.

CIRCULAR BUSINESS MODEL

In 2020, global textile consumption is estimated to be over 100 million per year. The environmental and social impact of this sector is an issue for the entire planet. Structuring sustainable supply chains today is a necessity for the fashion industry. In particular, the fast fashion industry, characterized by unbridled consumerism, has the greatest environmental and social impact. The model is called disposable because the focus is not on product quality but on efficiency and prices (De Giovanni, 2020).

In fact, this sector is based on a production system that makes intensive use of chemicals, large amounts of water and pesticides. The guidelines proposed by ISO 14000 should be followed. Among other things, this process is necessary in a context in which consumers pay more attention to sustainability issues and give them importance at the time of purchase (De Giovanni, 2021). Unlike a few years ago, when such initiatives came from the company in a top-down manner, it is now the market that demands certain social and environmental standards from the company. One solution to this problem is to manage the supply chain, or rather the value chain, in such a way as to use more natural resources, reduce CO2 emissions and achieve higher profit margins (Sacco and De Giovanni, 2018).

From this sector in particular, signals of a reversal of direction towards a more sustainable system at environmental and social level are arriving, also meeting the economic needs of companies, from a big Sweden company, H&M.

Circular economy and in particular Circular Business Model represent ways/opportunities that allow industries to operate in a more sustainable way in order to significantly reduce pollution and waste of resources. The circular business model (CBM), can also be defined as a business model that is based on circular economy practices, highlighted as 3R: reduce, reuse and recycle (Lieder and Rashid, 2016). However, CBM to take place, not only need to be implemented, but must also have the potential to gain market share from currently dominant linear business models (Bocken et al., 2018, Pal and Gander, 2018). With reference to the fashion retail value chain, the Circular Business Models operate within multiple areas, for example from the collection of used products to the sale of second-hand garments, rental, repair, redesign and reuse of recycled materials. (Pal and Gander, 2018). The

European Environment Agency (2019), as well as the Ellen MacArthur Foundation (2017), recognize that there are currently many initiatives in the fashion sector including fast fashion, CBM. But despite the willingness toward this model, in the Ellen MacArthur Foundation report they also write that "these efforts offer solutions and demonstrate promising progress in various areas, but are fragmented and often effective only on a small scale" (Ellen MacArthur Foundation, 2017, p.26). Among the critical elements is the decoupling between the company's overall business and CBM activity that Stål and Corvellec found in their study of Swedish apparel companies (2018). Another difficult element to manage is that customers not only constitute the demand side, but are also part of the supply side (Jalali et al., 2020), since they have to return their unwanted clothes to make them re-enter the resource cycle (Ki et al., 2020).

The literature on CBMs has highlighted four strategic approaches to facilitate the scalability of circular business models in the fashion industry. (Hultberg, E., & Pal, R. 2021):

1. 'Do it yourself'

Organizations that already have access to resources and capabilities within their business model can efficiently leverage them, through centralization, replication, standardization, and automation, to scale CBM tasks.

2. Divide the labor

Splitting work and resources between different organizations can be a way for organizations to leverage complementary resources and capabilities for a more cost-efficient process when expanding CBM activities outside of their areas of expertise. Therefore, finding the right partnership can be a source of scalability of the business model.

3. Absorb ideas and opportunities

Organizations can create or discover scalability opportunities by actively seeking insights from industry networks and customers, or by running pilots and experiments. The key is to act on these opportunities by absorbing them into the organization

4. *Create together*

Creating new ideas and technologies together in collaborative networks and communities is a source of learning, as well as a way to change the industry and thus also create better conditions to scale your CBM activities. However, to take advantage of this created opportunity, it is necessary to integrate it into the permanent business model.

In practical terms, however, these conceptual models should only be seen as a first step towards a more comprehensive tool for analyzing the scalability of CBMs but at the same time they could undoubtedly help companies analyze the appropriate actions for their circular initiatives.

Analysis of the Value, Drivers, Stakeholders, and Barriers.

Value Chain and Value Creation

By supply chain we mean the process of bringing a product or service to market, transferring it from the supplier to the customer. It is, therefore, a complex process that involves several professionals and activates numerous processes of the ecosystem-business: from the flow of raw materials related to production processes, up to the distribution logistics, which provides to get the purchased good to the customer (Vishkaie et al., 2021).

As a global fashion and design company, the impact of H&M's value chain on people, communities, economies and ecosystems is of greater importance and far-reaching, generating the phenomenon of "shared value". Shared value represents the set of policies and operating practices that strengthen the competitiveness of a company while improving the economic and social conditions of the communities in which it operates. (Porter and Kramer, 2011). Therefore, the Swedish company aims to ensure its own profitability and future growth through strategic investments and partnerships made to develop relationships with local communities (De Giovanni, 2022). The first impact generated comes from the new job opportunities created along with the growth of the brand. A second positive impact has been generated through participation in the Better Cotton Initiative, of which H&M has been a founding member since 2009. Precisely in line with the principle of shared value, this initiative allows the company to get the sustainable cotton it wants, albeit at a higher price, and allows farmers to increase their profitability and consequently strengthen their communities. To go beyond the boundaries of the corporate value chain, the "H&M Conscious Foundation" was created, an independent, global non-profit foundation that, through collaboration with various local partners, wants to improve conditions in the surrounding areas. The goal of H&M is to create value, through a customer base informed about environmental issues, about the problems that causes the fashion industry to the environment. The company has realized that people are increasingly sensitive and responsible towards these issues, sensing that the future trend will lead consumers to buy clothes that do not weigh on their consciences in terms of pollution or exploitation of labour (Duhaylongsod and De Giovanni, 2018). Following this awareness on the part of both the brand and its clients, H&M has made available data, reports and any information regarding its products or services, further enticing the consumer to become informed. H&M manages a dense network of commercial partners, which are carefully selected according to precise standards, because precisely the creation of value can only occur if everyone along the chain is aligned to the same strategy and objectives (De Giovanni and Folgiero, 2022).

H&M's drivers for the circular economy

To understand what can motivate all fashion stakeholders to adopt circular fashion (CF), the major drivers proposed in the literature were studied. "**Internal drivers**" included factors influencing CF adoption motivated by internal fashion stakeholders, while "**external drivers**" included motivational factors driven by external fashion stakeholders. Once divided, the drivers were further classified into different clusters:

- **Environmental concerns**: this cluster includes motivational factors related to the environment, such as the desire to protect the environment from the ecological footprint left by both internal and external fashion stakeholders.
- **Technology improvement**: represents technology drivers, such as the development of more resource-efficient production technologies.
- **Consumer Awareness**: discusses the key consumer motivating factors that drive fashion stakeholders to pursue CF.
- **Government policies**: represent the factors influenced by government announced rules or policies that motivate fashion stakeholders to adopt CF. For example. The EU Action Plan for the Circular Economy was adopted in March 2020 and provides guidelines for the life cycle of products so that they can stay in the market for the longest time.
- **Business opportunities**: this cluster includes business-related factors, such as new business opportunities in the CF market, that drive CF

adoption. For example, increased brand protection and loyalty causes H&M to differentiate itself from its competitors and capture a new branch of customers that it would not normally have captured. In fact, H&M is now able to attract customers usually averse to fast fashion.

- **Price competitiveness**: provides the factors related to cost or retail price competitiveness that motivate fashion companies to adopt CF. Waste management of suppliers and stores is very important for H&M. In 2020, 92% of the waste handled in H&M Group's distribution centres was reused or recycled.

Stakeholders and Barriers

Barriers to entry are those factors that make it difficult for a new company or start-up to enter a certain market. In fact, despite the fashion industry's efforts to implement Circular Fashion (CF), the organizational structure and contextual challenges inherent in the industry are still putting a strain on the sector. The issues are especially related to the highly fragmented organizational structure, consisting of a number of small units - design, sourcing, merchandising and marketing teams, which are highly interdependent on each other (Doppelt, 2003; Ha-Brookshire, 2017; Lozano, 2015), but also a long and complex supply chain for the industry to create CF (Maranesi and De Giovanni, 2020).

Indeed, H&M's supply chain is extremely fragmented with many stakeholders, internal (e.g., owners, managers, and employees) and external (e.g. government, recycling experts such as I:CO, agents, and consumers but also NGOs and IGOs), involved in the value creation process (Ha-Brookshire and others, 2017). The function of Circular Fashion is to reduce waste and pollution by extending the life of the products used, through reuse and recycling. Consequently, in order to reuse products and materials for a long life, H&M's circularity system needs the commitment of consumers in returning their unwanted clothes to the multinational company's stores so that these old products re-enter the supply chain for future recycling and reuse (De Giovanni and Zaccour, 2022). In this regard, as much as internal stakeholders have an influence on a firm's sustainability performance, this influence exerted by external stakeholders (e.g., consumers) plays an equally significant role in the firm's sustainable value creation (Jonker and Foster, 2002). In fact, among the key external stakeholders that help H&M create value are Partners such as the **Ellen MacArthur Foundation** who help promote the development of new ideas to improve the circular economy in retail. Recycling experts such as **I:CO** who offer new and innovative recycling opportunities through the provision of infrastructure to ensure that key raw materials from discarded fabrics enter a closed-loop production cycle.

We also analysed the barriers to value creation by H&M in developing this new circular model. The barriers were classified based on their relationships with internal and/or external stakeholders. "**Internal barriers**" include hindering factors found within companies or related to internal stakeholders. "**External barriers**" include barriers placed outside the companies tracked by external stakeholders. These barriers have been further classified clusters:

- *Technological and resource barriers*: to produce increasingly cutting-edge clothing collections with eco-sustainable materials, it is necessary to focus heavily on R&D, and very often it is possible to experience a gap with market competitors who exploit, either through themselves or through their suppliers, advanced technologies that enable increasingly green production.
- *Economic barriers*: some sustainable products are sold at a high price and perceived as an expensive luxury for some consumers. The benefit of having or consuming a green product is perceived as small and takes a long time before it can offset the expense of the products.
- *Knowledge barriers*: consumers tend to trust sustainability information provided or stated or claimed by scientists or environmental groups, but not by companies and not by businesses
- *Social barriers*: this cluster refers to low social awareness of EC and low public acceptance towards CF offerings as barriers to CF facilitation.
- *Government barriers*: this cluster refers to how the lack of supporting laws, or existing government policies, impose additional burdens on fashion stakeholders in their pursuit of CF.
- *Management Barriers*: this cluster includes lack of support from top management, or lack of cooperative/coherent management, among internal fashion stakeholders as barriers to CF implementation.

Network and Implications on Corporate Strategy

In 2003, H&M launched a collection of used clothes, starting a process of circularity in its business. However, it is necessary to analyze the objectives behind this circular economy model.

In 2011, it launched the "conscious" line, thus introducing a sustainable fashion line. Sustainable fashion supply chain includes eco-material

Table 1. Circular Economy model– H&M's objectives

To structure Circular supply chain	Help build supply chains that keep products in circulation and support circular production processes and material flows.
To create Circular products	Create products that are made to last, from safe, recycled and sustainably sourced materials and that can be repaired, reused, and remade multiple times.
To support Circular customer journeys	Make it easy for our customers to experience and engage in a circular fashion where products are used more, repaired, reused, and recycled.
Source: www.hmgroup.com	

preparation, sustainable manufacturing, green distribution, green retailing, and ethical consumers (Shen, 2014). H&M uses nearly 43% organic cotton which has minimal impact on the environment. H&M has a plan by 2030 to source all cotton from sustainable sources such as organic cotton, recycled cotton or Better Cotton. In 2017, the H&M group was ranked #1 by the Sustainable Cotton Ranking. "Organic cotton" is grown without chemical pesticides and fertilizers and does not contain genetically modified organisms (GMOs). "Recycled cotton" comes from old clothing and fabric scraps that are ground into fiber, new yarns and fabrics come from it and reuse it to make cotton. "Better cotton Initiative" through which 1 million farmers in 2015 received training to harvest cotton with less water and chemicals, H&M is the active member of this non-profit organization, in collaboration with partners such as WWF and Solidaridad. (Javed, 2020)

So, through these activities, H&M creates a true ecosystem where suppliers, partner companies and consumers collaborate proactively, creating four main streams, as you can see in the figure.

A true ecosystem is created in which suppliers, partner companies and consumers collaborate proactively, creating four main flows, as can be seen in the figure. H&M proposes to operate actively within all these processes, in particular through activities of:

- Products and customer offerings.
- Supply chain
- Non-commercial goods, such as packaging and items used in stores, offices and distribution centers

H&M does not personally produce the clothes it uses and always maintains a small inventory for reasons inherent to its business model that need not be

Figure 1. Ecosystem flows Source: www.hmgroup.com

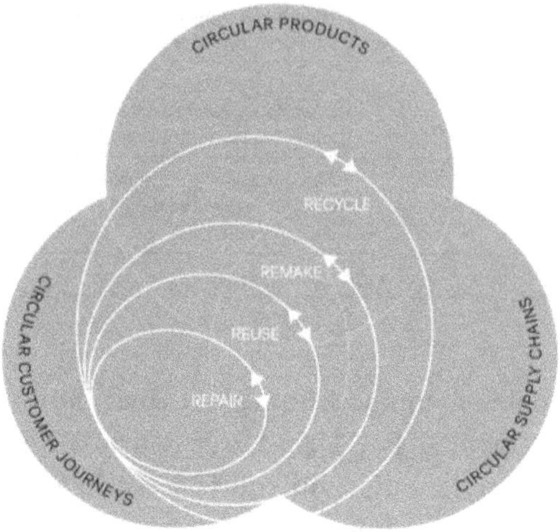

analysed here. Despite this, the logistics function is completely internalized. H&M is present in 74 states with over 5000 stores. The process of collecting used clothing takes place directly within H&M stores, which are periodically collected and redistributed through a partner company I:CO.

At this point, the used clothes are introduced into a sorting process, distinguishing the clothes according to their degree of wear and distributing them to processes of:

- Second-hand sales (returning to stores)
- Transformed into other products (reproposed)
- Recycled

H&M aims at sending zero waste from the organization to landfills. Recycled materials are popularly used at H&M. In addition to recycled cotton from textile remnant in production, H&M also uses recycled polyester, recycled polyamide, recycled plastic, and recycled wool in product lines. The advantages of using these recycled materials include saving energy and water, as well as lowering the greenhouse gap emission. To ensure the standard of sustainable materials, H&M is certified by independent third-party certification for each category (Shen, 2014).

It is useful to dwell on recycling, in fact, at this point the textile fibers are either reused in the textile sector creating a closed-loop or are destined

Figure 2. Circular economy network Source – www.ico-spirit.com

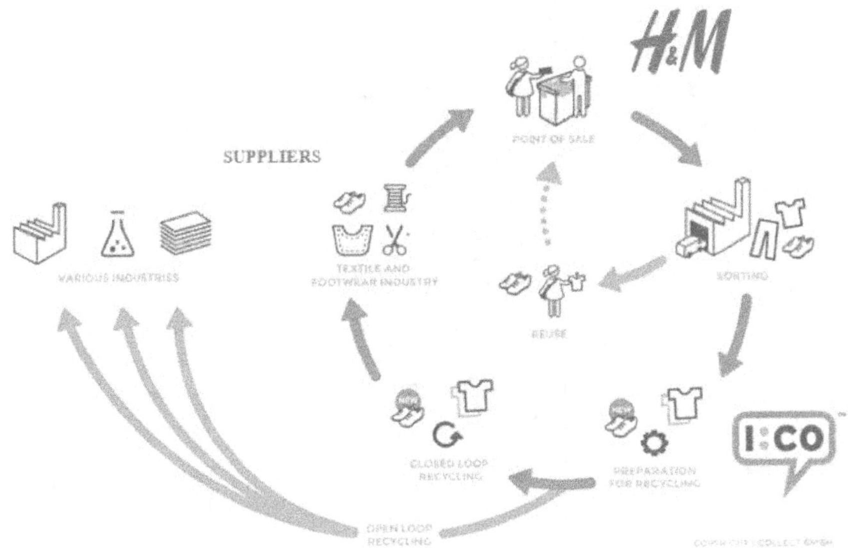

for other industries (most of the fibers are turned into insulation material for the automotive and construction industries). Furthermore, I:CO conducts “Circular Projects” with other partners to explore other recycling options.

When it comes to materials that fall into the closed loop, H&M has started a line of “conscious” products. They are made with at least 50% sustainably sourced materials such as organic cotton and recycled polyester - but many Conscious products contain much more than that. The only exception is recycled cotton, which can only account for 20% of a product.

However, H&M does not own the production of its own products and therefore needs to evaluate suppliers and their products. In the H&M group they use third-party life cycle assessment (LCA) data to assess the environmental impact of recycled materials and other sustainable sources. This includes individual LCAs and external material benchmarks based on LCA data, such as the Material Sustainability Index (MSI) from Sustainable Apparel Coalition (SAC). They encourage material manufacturers to contribute to the MSI by submitting their data and having their material analyzed in a transparent way. This helps everyone make more informed choices. Their more sustainable materials typically require the use of credible third-party certification systems. In this way, recycled products re-enter H&M’s cycle making the circular economy process effective.

Figure 3. Network structure

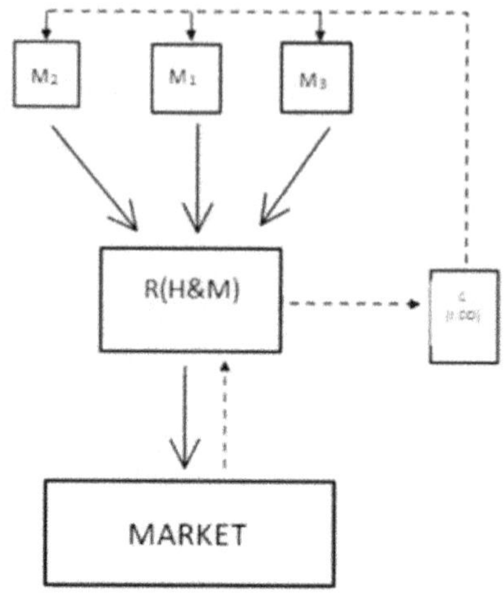

H&M thus finds itself offering three types of products: the classics, the second-hand ones and the conscious line. With those of second hand manages to re-evaluate products that would otherwise be garbage if not returned to the market, offering among other things a good product at a significantly lower price.

As for the costs of collecting clothes, these are limited simply to the granting of vouchers to customers. This represents the company's feedstock, which it will partly keep and partly sell to the next player in the circular process (I:CO). This in turn will collaborate as we have already seen with other players to revitalize these scraps.

To find the right partner in this delicate process, it is necessary to align shared values. To do this, H&M has chosen to collaborate with I:CO, a company that stands out for its mission: to act as a provider of innovative collection, certified sorting, reuse and recycling services. In this way the company can close the closed-loop previously mentioned.

I:CO is present in Germany, USA, Japan, China, France and in 60 other countries including Italy. You can identify I:CO as a great company by looking at the numbers. To date, the company has collected more than 90 thousand tons of used clothes and in 2017 alone reached a volume of 22 thousand tons. It collaborates with over 40 retail partners, thus ensuring that the circular industry has enough input to achieve economies of scale. I:CO's corporate benefits include:

- An innovative and cost-effective in-store collection concept to engage consumers, reward, drive store traffic and sales, and reach worldwide. Certified sorting and recycling of collected clothing and footwear allows us to provide a more sustainable solution at the end of the product's end-of-life.
- A convenient service at the point of sale in retail stores that accommodates clothing and shoes that are no longer needed and rewards shoppers. This enables our customers to use textiles more efficiently and change their lives and society for the better.
- I:CO recovery system helps reduce textile waste. Reusing and recycling clothing and footwear saves resources and helps to recycle valuable raw materials, ultimately helping to protect the environment. This forms the basis of a circular economy in the textile industry.

After I:CO's collection process, the feedstock enters a recycling process operated by SOEX the parent company of I:CO. Used fabrics that are no longer available make up about 30% of the goods SOEX receives. Simply disposing of these items is a waste of valuable resources. To combat this waste, SOEX sells some of its products to the textile industry. Other parts are shredded in recycling facilities and turned into secondary raw materials (mixtures of recycled materials).

At this point, having obtained the textile fibers from the recycling process, part of them will return to the clothing industry. At this point, H&M proceeds to an intense process of selection and subsequent analysis of suppliers. The results can be found in the GRI report, under items 103-1/2/3 and 308-1

First of all, it should be noted that H&M, before entering into any business relationship with any other company, signs a code of ethics with them, in order to ensure loyalty, reliability and collaboration. They monitor compliance using a system of risk-based audits and through insights into performance rather than audits alone, creating a positive feedback loop to empower constant improvement. This includes minimum requirements for new entrants, a self-assessment process, improved capabilities with a focus on renewable energy and sustainable processes and, finally, a proactive approach to incidents.

In this way, H&M, even not having a direct control over the activities of suppliers, is able to guarantee to the final customer eco-friendly and eco-sustainable products. Therefore, the company plays a crucial and fundamental role in the whole circular process of the industry. Customers, too, are considered an active part of the circular economy cycle, indeed, they are understood as the first promoters of the reuse cycle. In fact, as we have seen, H&M starts the process in their stores by placing garment collection totems, where customers can hand in their garments and in return receive a coupon to be reused within the H&M store. For every bag of clothing and home textiles that are delivered, the company will give a coupon worth €5 to be used on a minimum spend of €40. In addition, the totems are strategically placed inside the stores in order to ensure a more effective touch point towards customers, just look at I:CO that with all retail partners acts in this way. Consistent with its ethics, H&M also donates €0.02 for every kilogram of clothing collected to a local charity chosen by the company.

So, this process allows the customer to be more and more enticed to return the clothes to be reused, obtaining more and more coupons and thus becoming a best customer for the company. In addition to being at the base of the circular process, customers also operate during the lifecycle of the product, in fact already on the H&M website it is possible to find tips to extend the life of products or reinvent their use thanks to the "H&M Take Care" program.

H&M has based its business model for over 70 years mainly on 2 factors: customer-centric approach and customer value creation. To do this, over the years it has always strived to ensure affordable prices for many people around the world. The need, in today's context, to adopt a proactive approach has also affected H&M, which, while always maintaining high performance in the social sphere, in recent years has made great strides from the environmental point of view driven by the creation of shared value. Recognizing this not only as a challenge but also as an enormous opportunity, the company has invested heavily in innovation, making it a distinctive feature of its business. If today H&M counts in its network partners of international caliber and suppliers of services and goods at the forefront, this is due to the credibility, empirically proven, of its commitment in the social and environmental field. Today, more than ever, the company is committed to making a difference for employees, local communities and, most importantly, planet earth. Driven by this objective and by the previously mentioned objectives of "To structure Circular supply chain", "To create Circular products", "To support Circular customer journeys", the company has increasingly opened to the external environment, encouraging a collaborative approach with partners and customers. In fact, the company has created numerous platforms to get in touch with these parties and, in a certain sense, involve them in the "newborn" process of circular economy. In recent years, H&M has been committed to ensuring the empowerment of these external stakeholders. All this takes place in a perfectly eco-sustainable and energy transaction context.

But getting into the specifics, let's now look at how H&M actually moved over time to make this possible. In the face of global challenges, H&M has not been unprepared, indeed it has created a vision based almost entirely on sustainability. It seems clear, even from what has been said so far, that the company aims to lead a change towards circular and environmentally positive fashion. Three of the ambitions outlined in the company's vision is:

- Being a fair and solidarity company aiming at a fair access to resources and sustainable solutions;
- Tackling the environmental crisis with every effort, aiming above all at product circularity;
- The social and environmental impact of the company must be as a real example for customers and all stakeholders concerned.

What H&M has developed in recent years is a strategy called the "Change-Making Program", a set of innovative iterative tools and methods that enable the organization to achieve corporate and KPI goals.

H&M is succeeding in achieving sustainability goals in recent years through strategic plans, goals, actions and also through the facilitation of the company's strong culture and values.

The last competitive advantage of H&M is that thanks to its principles and strategies, it can actively involve all partners involved in their value chain, generating a homogeneous vision of sustainability but still allowing to maintain a unique identity.

In this way H&M has been able to record important results in relation to the numerous issues mentioned.

- Circularity: 76% (54% in 2019).
- Recycled or sustainably sourced materials: 73% (57% in 2019).
- Climate positive: 62% (45% in 2019).
- Transparency: 80% (58% in 2019).
- Innovation: 78% (57% in 2019).
- Inclusion and diversity: 47% (39% in 2019).
- Fair jobs: 43% (35% in 2019)

"At H&M Group, we are continually working to make our business model more sustainable to secure long-term growth and, at the same time, make a positive contribution to the world. To us, this means creating decent and meaningful

jobs and promoting inclusiveness and diversity. It means becoming climate positive throughout our entire value chain by 2040, as well as achieving circularity. Accelerating our agenda to become a circular business is one of our key focuses. We don't have all the answers ourselves, but by teaming up with – and investing in – pioneering companies that develop groundbreaking technologies, such as Re:newcell or Infinite Fiber Company, we can scale innovations and reinvent fashion together. I'm confident that our close collaborations with organizations such as the Ellen McArthur Foundation, WWF, IndustriAll, the UN Global Compact and many others will continue to inspire us, help us find even more solutions to transform ourselves and share best practice with both partners and competitors".

These are the most salient parts of the speech of the company's CEO, Helena Helmersson, reported in the Sustainability Report 2019, which make us understand how the H&M group has been at the forefront of implementing circular business models for years through a series of initiatives that we will analyze below:

3R: Reduce, Reuse, Recycle

As we mentioned before, the H&M Group has been collaborating since 2013 with I:CO. they had to re-design all the brand's stores setting up special garment collection baskets where customers can leave the clothes they do not use by obtaining discount coupons for future purchases. The company that manages H&M's collection program sorts all used fabrics into three categories based on how they are then reintegrated into a new lifecycle:

- *Re-garments*: garments that are still in a position to be worn are sold as second-hand clothes.
- *Reused*: old clothes and fabrics are transformed into other products such as, for example, cleaning rags.
- *Recycled*: the remaining fabrics will be transformed into textile fibers for the production of other garments or will be used for other purposes such as, for example, insulating material.

Another initiative proposed by H&M to solve the problem of *overconsumption* in the Fast Fashion sector is the "Take Care" program. Through this program, H&M emphasizes once again that its goal is not to make the most of profit by enticing consumers to buy and throw away as much as possible, but rather aims to 'create' consumers who are more aware of the damage of the Fast Fashion industry and more sensitive to environmental issues.

Actions for Less Pollution

The H&M group, which considers attention to the planet part of its competitive advantage, seeks to adopt an increasingly "green" policy. In order to improve its behaviors, the H&M company has identified a set of KPIs (Key Performance Indicators) whose value is monitored and made public annually within the sustainability report.

In particular, H&M notes the performance of the following KPIs:

- Annual percentage change in the CO2 emissions.
- Percentage change in the intensity of electricity use (consumption of kwh per square meter of surface and per hour of opening to the public) using the 2016 figure as a basis.
- Percentage of energy consumption from renewable sources in business operations.
- Percentage of recycled water compared to total consumption.
- Percentage of sites with availability of efficient plants compared to water consumption.
- Percentage of recycled (or sustainably sourced) materials in total materials used in commercial products.
- Percentage of cotton recycled (or from sustainable sources).
- Tons of garments collected through the specific collection initiative.
- Percentage of "stores" with recycling systems of the main types of waste.

With regard to means of transport, the H&M group, with the aim of having the least possible impact on the environment, has taken part in two interesting collaborations and coalitions. The first is the collaboration with the company Maersk, a leading transport company in its sector, which has invented a biofuel mixture that can replace fossil fuels and reduce emissions from transport by ship. The second initiative is the coalition "The Pathways" with Scania, E. ON and Siemens, with the aim of accelerating the reduction of the use of fossil fuels in heavy commercial transport, through better logistics, the use of electric vehicles and ecological fuels.

Finally, the H&M group has also committed itself with reference to the chemicals used during all stages of processing, collaborating with suppliers so that in the washing, dyeing and printing phases of the garments they meet even more stringent requirements than the normal regulations imposed on industries. H&M was one of the first companies to have drawn up, as early as 1995, a list of restrictions on chemicals allowed in processing, continuously updated. To ensure the choice of the best chemicals, it has adopted a method called "Screened Chemistry" that evaluates and certifies chemicals based on their impact on human health and the environment.

Circulose and Vegea

H&M was the first retailer to use Circulose, a new material made entirely from discarded fabrics, patented by Renewcell, a company financed by H&M itself. Circulose is a natural material that is obtained starting from the cellulose present in the fabrics used, with which new yarns are made with the typical characteristics of vegetable fibers. This material has the advantage both of being able to recycle garments over and over again without reducing their quality, and of being able to produce new garments without the need for raw materials. Another environmentally friendly material used by H&M is "Vegea". This fabric is produced thanks to a special treatment of the fibers and oils present in the 'vinaccia', a natural material consisting of the waste of wine production such as seeds and skins.

Circular Economy System and Performance

H&M's Corporate Sustainability department is divided into four groups (Social, Environmental, Production and Relational) that work in synergy, with the unique goal of bringing the company to be a totally circular system, so that it can meet the needs of today's customers. Thanks to the work of each department, it is possible to carry out an in-depth analysis of the company's performance to date. Currently from the results of H&M's 2020 Sustainability Report, the total assortment of products reused (including donations to charity) or recycled due to defective products is 0.32% (0.41% in 2019). Total Assortment of destroyed products were 0.03% (0.03% in 2019).

The Waste managed in the distribution centers of the H&M Group amounted to 50,170 tons, of which 92% was recycled or reused, in addition, the fabrics collected in the stores of the H&M Group amounted to 18,800 tons (29,005 in 2019), equivalent to more than 94 million T-shirts. Despite the excellent results, H&M is increasingly looking to the future, in fact, an ambitious new materials target has been set, aiming to use 30% recycled material by 2025. This is supported by innovations in the recycling of post-consumer fabrics without loss of quality, such as Monki's collection in collaboration with HKRITA's Green Machine. The H&M Group has also developed a multi-brand packaging system with certified paper bags, which have been introduced to customers at COS, ARKET, Monki and Weekday and in selected H&M brand markets and will be further implemented in the coming months.

POSSIBLE IMPLICATIONS FOR SMART CITIES

Smart cities can be defined as: "A city connecting the physical infrastructure, the IT infrastructure, the social infrastructure, and the business infrastructure to leverage the collective intelligence of the city. [...]. In a Smarter City, the traditional concept of a physical city infrastructure is extended to a virtual city infrastructure, an integrated framework that will allow cities to gather, integrate, analyze, optimize, and make decisions based on detailed operational data". (Harrison et al., 2010).

They aim to improve the quality not only of infrastructure, but also and above all of governance and the lives of citizens. It is not enough to infuse technologies in all subsystems of a city - e.g., transport, energy, health - but it is indeed necessary to consider the city as an organic system, where the main connectors are the people who live there (R.M. Kanter et al., 2009). The pillars of smart cities are environmental, social and economic-financial sustainability on which the smart city must be based in order to be successful.

In a smart waste perspective, the current management of textile waste is not sustainable, the European Parliament has intervened with a directive on waste, which obliges EU countries to provide separate collection of textiles by 2025. The Commission's new strategy also includes measures to support circular materials and production processes, to combat the presence of hazardous chemicals and to help consumers choose sustainable textile products.

There was a focus on smart waste, as H&M's clothing recycling initiatives could be well placed in a smart city. In addition, smart people can be identified with the H&M customer, who is attentive to issues of sustainability and circular economy. All the Swedish company's initiatives have a common thread with the domains identified by (Giffinger et al., 2020).

H&M has developed a recycling system: Looop, which, directly in-store, transforms old clothing into new. An unwanted garment is fed into the machine to be cleaned, shredded and spun into yarn. After about five hours, a new knitted garment is made from the garment fed into the machine. The system in question does not use water or chemicals, thus having a significantly lower environmental impact than if the garments were produced from scratch.

The innovative technology behind it was developed by the Hong Kong Research institute of textiles and apparel (HKRITA) in collaboration with the non-profit H&M foundation. Currently, Looop is the only in-store (Stockholm) recycling machine, but HKRITA will license the technology on a large scale to help the entire industry become more circular. Loop could fit well into a smart context as it respects the cardinal principles of smart cities and aims to deviate from fast fashion as much as possible while reducing environmental impact.

CONCLUSION

Nowadays, the fashion industry, and in particular the fast fashion one, has a significant impact on the environment and society. Consumer demands are no longer limited to the mere product but require a perfectly sustainable system at the basis of processes. Fashion companies, and especially the largest ones, must move towards this by circularizing their entire business model. A pioneer in this field is undoubtedly the Swedish giant H&M, which for several years now has been launching various sustainable management initiatives and has become a leader within the network. Other companies in the sector are beginning to emulate H&M and the emergence of the smart cities phenomenon is opening up ways for developing this model at public level. However, the environmental and social impact of the sector is still significant, and the question now is whether this model, with technological development, will be a definitive solution to the main sustainability issues.

REFERENCES

Battaglia, M., Testa, F., Bianchi, L., Iraldo, F., & Frey, M. (2014). Corporate social responsibility and competitiveness within SMEs of the fashion industry: Evidence from Italy and France. *Sustainability*, *6*(2), 872–893. doi:10.3390u6020872

Bocken, N. M., Schuit, C. S., & Kraaijenhagen, C. (2018). Experimenting with a circular business model: Lessons from eight cases. *Environmental Innovation and Societal Transitions*, *28*, 79–95. doi:10.1016/j.eist.2018.02.001

Chan, T. Y., & Wong, C. W. (2012). The consumption side of sustainable fashion supply chain: Understanding fashion consumer eco-fashion consumption decision. *Journal of Fashion Marketing and Management*, *16*(2), 193–215. doi:10.1108/13612021211222824

De Brito, M. P., Carbone, V., & Blanquart, C. M. (2008). Towards a sustainable fashion retail supply chain in Europe: Organisation and performance. *International Journal of Production Economics*, *114*(2), 534–553. doi:10.1016/j.ijpe.2007.06.012

De Giovanni, P. (2016). Coordination in a distribution channel with decisions on the nature of incentives and share-dependency on pricing. *The Journal of the Operational Research Society*, *67*(8), 1034–1049. doi:10.1057/jors.2015.118

De Giovanni, P. (2019). When feature-based production capabilities challenge operations: Drivers, moderators, and performance. *International Journal of Operations & Production Management*, *40*(2), 221–242. doi:10.1108/IJOPM-04-2019-0309

De Giovanni, P. (2020). An optimal control model with defective products and goodwill damages. *Annals of Operations Research*, *289*(2), 419–430. doi:10.100710479-019-03176-4

De Giovanni, P. (2021). *Dynamic quality models and games in digital supply chains*. Springer Nature. doi:10.1007/978-3-030-66537-1

De Giovanni, P. (2022). Leveraging the circular economy with a closed-loop supply chain and a reverse omnichannel using blockchain technology and incentives. *International Journal of Operations & Production Management*, *42*(7), 959–994. doi:10.1108/IJOPM-07-2021-0445

De Giovanni, P., & Folgiero, P. (2022). *Strategies for the Circular Economy*. Circular Districts and Networks. Giappichelli-Routledge.

De Giovanni, P., & Vinzi, V. E. (2014). The benefits of a monitoring strategy for firms subject to the Emissions Trading System. *Transportation Research Part D, Transport and Environment*, *33*, 220–233. doi:10.1016/j.trd.2014.06.008

De Giovanni, P., & Zaccour, G. (2022). A selective survey of game-theoretic models of closed-loop supply chains. *Annals of Operations Research*, *314*(1), 1–40. doi:10.100710479-021-04483-5

Duhaylongsod, J. B., & De Giovanni, P. (2018). The impact of innovation strategies on the relationship between supplier integration and operational performance. *International Journal of Physical Distribution & Logistics Management*, *49*(2), 156–177. doi:10.1108/IJPDLM-09-2017-0269

Giffinger, R., Fertner, C., Kramar, H., Kalasek, R., Pichler-Milanovic, N., & Meijers, E. (2007). *Smart cities. Ranking of European medium-sized cities.* Final Report, Centre of Regional Science, Vienna UT.

Harrison, C., Eckman, B., Hamilton, R., Hartswick, P., Kalagnanam, J., Paraszczak, J., & Williams, P. (2010). Foundations for smarter cities. *IBM Journal of Research and Development*, *54*(4), 1–16. doi:10.1147/JRD.2010.2048257

Hultberg, E., & Pal, R. (2021). Lessons on business model scalability for circular economy in the fashion retail value chain: Towards a conceptual model. *Sustainable Production and Consumption*, *28*, 686–698. doi:10.1016/j.spc.2021.06.033

Jalali, H., Ansaripoor, A. H., & De Giovanni, P. (2020). Closed-loop supply chains with complementary products. *International Journal of Production Economics*, *229*, 107757. doi:10.1016/j.ijpe.2020.107757

Javed, T., Yang, J., Gilal, W. G., & Gilal, N. G. (2020). The sustainability claims' impact on the consumer's green perception and behavioral intention: A case study of H&M. *Advances in Management and Applied Economics*, *10*(2), 1–22.

Ki, C. W., Chong, S. M., & Ha-Brookshire, J. E. (2020). How fashion can achieve sustainable development through a circular economy and stakeholder engagement: A systematic literature review. *Corporate Social Responsibility and Environmental Management*, *27*(6), 2401–2424. doi:10.1002/csr.1970

Lieder, M., & Rashid, A. (2016). Towards circular economy implementation: A comprehensive review in context of manufacturing industry. *Journal of Cleaner Production*, *115*, 36–51. doi:10.1016/j.jclepro.2015.12.042

Maranesi, C., & De Giovanni, P. (2020). Modern circular economy: Corporate strategy, supply chain, and industrial symbiosis. *Sustainability, 12*(22), 9383. doi:10.3390u12229383

Moss Kanter, R., & Litow, S. S. (2009). *Informed and interconnected: A manifesto for smarter cities.* Harvard Business School General Management Unit Working Paper (09-141).

Pal, R. (2017). Value creation through reverse logistics in used clothing networks. *International Journal of Logistics Management, 28*(3), 864–906. doi:10.1108/IJLM-11-2016-0272

Sacco, A., & De Giovanni, P. (2019). Channel coordination with a manufacturer controlling the price and the effect of competition. *Journal of Business Research, 96*, 97–114. doi:10.1016/j.jbusres.2018.09.001

Sandberg, E., Pal, R., & Hemilä, J. (2018). Exploring value creation and appropriation in the reverse clothing supply chain. *International Journal of Logistics Management, 29*(1), 90–109. doi:10.1108/IJLM-10-2016-0241

Shen, B. (2014). Sustainable fashion supply chain: Lessons from H&M. *Sustainability, 6*(9), 6236–6249. doi:10.3390u6096236

Shen, B., Zheng, J. H., Chow, P. S., & Chow, K. Y. (2014). Perception of fashion sustainability in online community. *Journal of the Textile Institute, 105*(9), 971–979. doi:10.1080/00405000.2013.866334

Stål, H. I., & Corvellec, H. (2018). A decoupling perspective on circular business model implementation: Illustrations from Swedish apparel. *Journal of Cleaner Production, 171*, 630–643. doi:10.1016/j.jclepro.2017.09.249

Vishkaei, B. M., Fathi, M., Khakifirooz, M., & De Giovanni, P. (2021). Bi-objective optimization for customers' satisfaction improvement in a Public Bicycle Sharing System. *Computers & Industrial Engineering, 161*, 107587. doi:10.1016/j.cie.2021.107587

Chapter 9
Circular Economy in the Textile Industry

Pietro De Giovanni
https://orcid.org/0000-0002-1247-4807
Luiss University, Italy & Luiss-X.ITE Research Center, Italy

EXECUTIVE SUMMARY

This chapter elaborates on the application of circular economy systems in the textile industry, which is historically influenced by seasonality effects that induce a search for circular solutions independent of environmental constraints and regulations. This chapter shows that circular economy systems can create business opportunities as well as important cannibalization effects when consumers are confronted with new and used textile goods. Hence, it highlights how the use of digital technologies can help to support the mitigation of such issues..

INTRODUCTION

This business case focuses on analysing Patagonia Inc., an American textile firm that changed the world of fashion in the last decades. This firm has built a clothing empire through its love for the environment and a passion for crafting quality goods that enhance humanity's relationship with it. Patagonia's biggest legacy may not just be its clothing, but its entire brand philosophy. Indeed, buying Patagonia's clothes means to embrace a sustainable and innovative lifestyle. Environmental and social sustainability, technological innovation and comfort are the main features of this brand, which has been piloting the

DOI: 10.4018/978-1-6684-5001-7.ch009

textile sector into this new ethical paradigm (De Giovanni, 2019). Moreover, Patagonia is involved in various partnerships with research centres to address sustainable farming matters, thanks to a fully dedicated division called Patagonia Provision. This is a clear example of the all-around commitment of the brand to the health of our planet.

The aim of the present work is to analyse Patagonia's Worn Wear initiative, which puts in practice the firm's beliefs in environmental and social sustainability, offering a true example of a functioning circular business model. The project was launched in 2013 and has become integral part of Patagonia's success on the market, highlighting in counter-intuitive way the quality and durability of the brand's garments and attracting new talents and conscious consumers every year (De Giovanni, 2020). Indeed, Patagonia's philosophy is strongly based on community engagement and, together with valuable partnerships, the brand was able to build a real success story.

The business case retraces the firm's journey to success, addressing the birth of Worn Wear and how it affects the business values, strategy, and economic performance. The first section is dedicated to the circular economy network, followed by the analysis of the drivers, values and stakeholders. Then, the chapter addresses how circular economy affects Patagonia's corporate strategy, giving insights about the Worn Wear initiative and other projects in which the brand is involved. Furthermore, the present work acknowledges the connection between circular economy and performance, together with a pragmatic analysis of two similar brands on the market: The North Face and Save the Duck. Eventually, the last sections include the circular economy developments, business transformations and possible implications for smart cities. To have a general idea of the company's strategy, the SWOT analysis proposed by Shastri (2021) is displayed in Figure 1.

CIRCULAR ECONOMY NETWORK

From the motto *"If It's Broken, Fix It"*, Patagonia mother brand gave born to Worn Wear. Worn Wear defines itself as "a community of climbers attracted by the most aesthetic line and motivated to protect the places where they climb".

Patagonia Worn Wear works as proper *hub* where several actors are involved aiming at extending the life of products. Multiple players are essential for the functionality of the platform; indeed, in our case study are extra steps to

Figure 1. The SWOT analysis of Patagonia

the buying and selling process. Repair, Reuse and Recycling are the three milestones around which the platform is built. If a product is totally worn out, they take it and recycle it. If it needs to be repaired, they have a mechanism to repair it. If it can be sold but they don't want it anymore, another tool is eligible to do that.

As already mentioned, Worn Wear operates as proper network of players with a common final aim: extend the life of products. Every and each action made by the players is essential for the functionality of this circle. Here, it follows a description of the main actions the platform carries out and the major relative actors in each of those steps.

When it comes to get back the "fully used" products into the loop, costumers are the actors playing the first game. Indeed, the entire platform starts acting once the costumers interact with it putting the garb back in the system. One of the most responsible things a consumer can do for the company is to keep the products in use as long as possible. Once the product reaches its highest level of usage, the consumer drops off the item and receive credit in stores. Given that this trend is peculiar and mainly spread among youngest generation, who Worn Wear is trying to attract the most with its campaign are younger consumers. Patagonia, as leading role in this "circle" (i.e., the manufacturer) gives store credit towards next purchase, on a used or new garment in exchange for getting back its own products in adequate conditions. In this regard, Patagonia also provide for the main infrastructures essential for this process (Duhaylongsod and De Giovanni, 2018).

Patagonia also is committed to assume the responsibility of products after selling. It derives several partners involved in the repairing process. They hold one of the biggest repair centres in North America and are supported by more than 70 other repair centres around the world, which allow Patagonia to repair more than 100,000 items each year. Moreover, they offer this service through virtual repair guides. Indeed, Patagonia mainly partnerships with the fellow repair junkies at iFixit. The latter shows costumers all the steps on how to repair Patagonia clothing. Finally, they run mobile repair stations, which travel around North America to offer low-cost garment fixes.

When items are totally worn by consumers - meaning that they do not reflect the standard to be rejected again in the loop, and in turn cannot be repaired or reused, - it comes the Recycle phase. Owing to Patagonia clothing is so durable and made up of mixed materials, it is difficult to find recyclers that are willing to accept postconsumer wastes. As whole, Patagonia's recycling process foresees two main options: collected Patagonia clothing from customers providing for recycling bins into global stores upon consumer's request; and occasionally partnerships with companies to create upcycled products and inject them into the resale phase.

At this regard, what worth to underline is that Patagonia is trying to anticipate a growing retailer trend through *Worn Wear*. In fact, the future approach in the panorama of resale is "brands wanting to own that customer", and Patagonia is trying to catch it with their new philosophy. The start-up has seen a 300% to 400% growth year-over-year in partnership with retailers, highlighting retail demand.

Circular Supply Chain

One important development that Patagonia worked out recently is the implementation of a Circular Supply Chain. Figure 2 reports an example of a circular supply chain (De Giovanni and Folgiero, 2022), indicating the actors intervening in the management of forward and reverse flows and showing the possible circular economy options available.

Circular Supply Chain Management can be defined as "the integration of circular thinking into the management of the supply chain and its surrounding

Figure 2. A typical Circular Supply Chain in the textile industry

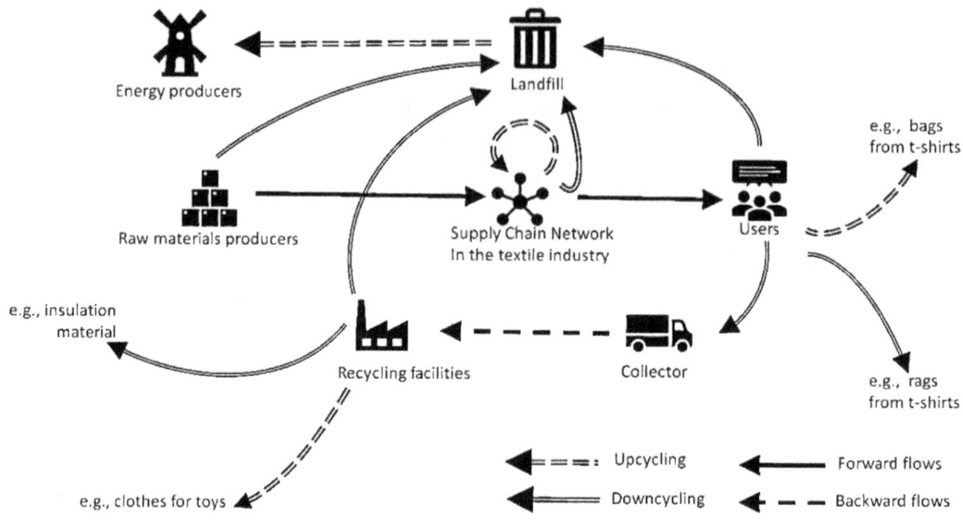

industrial and natural ecosystems. It systematically restores technical materials and regenerates biological materials toward a zero-waste vision through system-wide innovation in business models and supply chain functions from product/service design to end-of-life and waste management, involving all stakeholders in a product/service lifecycle including parts/product manufacturers, service providers, consumers, and users" (Farooque and Zhang,2018).

From a theoretical point of view, the aim of a CSCM is to generate zero waste through the collaboration between the producer's supply chain and secondary chains thanks to which a company can restore and regenerate resources. This aim can be achieved only thanks a Cross-Sector Collaboration that is based on industrial symbiosis. Hofmann and Jaeger-Erben (2020) gives a definition of Cross-Sector collaboration that is when "independent actors from different sectors collaborate and negotiate to share their resources and develop their core capabilities"; usually it is a long-term relationship that is based on risks sharing and common goals. An example of this phenomena can be The Circle-House-Project that is a cross-sectoral collaborative project aimed at creating circular economy in the Danish construction sector; this sector is one of the most resource- intensive because consume a lot of raw materials, energy and water and it also include a multitude of suppliers. The aim of The Circle-House-Project is to implement the diffusion of sustainable practices within this sector and include the largest possible number of participants "right from the design phase, to ensure that the knowledge created stretches as far as possible" (Köhler et al., 2021).

Another example of Cross-Sector collaboration is the one between Volvo and Battery Loop that are working together to give a second life to the batteries, used by Volvo in their cars, to create a solar-powered energy storage system.

CSCM has its foundation on three elements: natural environment, society, and economic performance at a broader level that represents the three aspects of the organizational sustainability. By modifying the weight of these three factors, various concepts have been developed in literature:

- Sustainable Supply Chain Management®It is the management of material, information and capital flows as well as cooperation among companies along the supply chain while taking goals from all three dimensions of the triple bottom line. (Seuring and Muller, 2008).
- Green Supply Chain Management (GSCM)® It represents the integration of environmental thinking into supply chain management, including product design, material sourcing and selection, manufacturing processes, delivery of the final product to the consumers as well as end-of-life management of the product after its useful life. The aim of GSCM is to mitigate the environmental impact of not only the supply chain but also the entire organization. (Srivastava, 2007).
- Environmental Supply Chain Management®It represents the set of objectives, plans and management systems that determine manufacturing's position and responsiveness to environmental issues and regulation (Zsidisin and Siferd, 2000).
- Closed Loop Supply Chain® focuses on product returns and essentially combines the traditional supply chain with the reverse logistic, after that the product reached its end-of-life or end-of-use lifecycle. (Guide and Van Wassenhove, 2006).

In summary, CSCM improve the concept of Sustainable Supply Chain Management and Green Supply Chain Management because apply the circular thinking in all supply chain phases and functions and moreover is applicable both to manufacturing product and to services.

Differently to Closed Loop Supply Chain, CSC includes a secondary supply chain that not only guarantee back flow but also create an added value form waste thanks to the collaboration with other organizations which may belong to a different sector.

During the last years, firms had embraced new business models based on the re-use and upcycling of raw materials with the implementation of sustainable practices, to face the climate changes (Genovese et al., 2017).

To achieve better results and develop a circular supply chain, firms should introduce more sustainable practices and collaborate with external stakeholders to guarantee the reverse flow and the creation of an added value for the products (Del Giudice et al., 2020). Consequently, the implementation of a greener supply chain can be translated, in most of the cases, into important financial and commercial benefits, in fact, in most of the cases, companies have a reduction of the operating costs, a stronger brand and a greater ease of access to resources (Gideon et al.,2020).

This interest in circularity is also shared with customers who are very concerned about the impact of their consumption experience and want to melt the economic and personal wellness with the environmental and social ones (Tseng et al., 2019).

The introduction of the sustainability concept into the Supply Chain Management has been very discussed in literature and in practice. Organizations are driven by various factors in adopting sustainable practices into their supply chain that can be classified in pressures, triggers, enablers, and drivers.

These factors are, first of all, pressures because push companies toward the application of specific sustainability measures (Caniato et al., 2012) but also drivers because motivate organizations in implementing CSC practices (Kösal et al., 2017). These elements are defined also triggers because push organizations to introduce new sustainable practices throughout the supply chain (Saeed, 2017).

Another important categorization in literature is the division in primary and secondary factors, depending on their contribution to introduce sustainable practices (Saeed, 2017). Primary factors are, for example, pressures from shareholders, suppliers, or employees so that have a direct influence on the supply chain and on the organization, in general. While the secondary drivers are pressures related to reputation, image, and NGOs so have an indirect pressure.

Analysis of the Value, Drivers, Stakeholders, and Barriers

Patagonia Inc. has four main values. The first one is *building the best product*. Among the most direct ways they can limit ecological impacts is with goods that last for generations or can be recycled so the materials in them remain in use. Making the best product matters for saving the planet. The second one is *causing no unnecessary harm.* They know that their business activity is part of the problem. They work steadily to change their business practices and share what they have learned. But they recognise that this is not enough. They seek not only to do less harm, but better (De Giovanni, 2016). The third main values is *use business to protect nature*. The challenges they face as a society require leadership. They embrace risk and act to protect and restore the stability, integrity and beauty of the web of life. The fourth is *not bound by convention.* Their success lies in developing new ways to do things.

Being in line with the values of the company, the Worn Wear start-up embodies perfectly all the values of the company. Indeed, the value proposition of Worn Wear is to give a valuable clothing reuse option for Patagonia customers. In this way, they simultaneously increase their consumers' engagement, enhance their brand, save costs (environmental, production and labor) and generate a new revenue on repaired products. Furthermore, in this process there is also a value creation: consumers can trade in their used products, receive a trade-in price, and Patagonia will give up to $100 in vouchers to be used online or in their stores.

The principal stakeholders of this initiative are the customers and Patagonia. Indeed, the relationship between the buyers and the sellers is a key aspect for this case. The customers need to prefer a second-hand cloth respect to a new one and on the other hand the firm have to guarantee a high quality and efficient service to its customers. Furthermore, the partnerships with reparation centres play a fundamental role in the Worn Wear system. Eventually, the public authorities and the waste managers/recyclers can be considered stakeholders because they do not have to manage the waste that otherwise Patagonia would produce without this initiative.

One of the barriers to the Worn Wear initiative are the customers themselves, because they might take advantage of this service even if they do not need it. However, Patagonia has a high customers' engagement that avoids for the majority this free riding behaviours. Indeed, the company has a precise and structured customer base that is embedded in its values. Furthermore, another barrier are the infrastructures to implement the initiative, because they need to create a solid network of shops where the customers can use the service. Furthermore, another barrier could be represented by the logistic of the infrastructure, since to repair and recycle the items, Worn Wear has to rely on different partnerships. The main issue is that Patagonia has to outsource these processes, with following costs and environmental impact

Table 1. Value proposition, stakeholders, drivers, and barriers

VALUE PROPOSITION	STAKEHOLDERS	DRIVERS	BARRIERS
Give a valuable clothing reuse option for Patagonia customers	Customers	Customers' engagement	Customers' behavior
	Patagonia	Brand enhancement	
	Waste managers and recyclers	Cost savings (production, environmental, labor)	
	Public authorities	Revenue generation	
	Reparation centers		

Indeed, once that the items are collected, in accordance with their characteristics, they can be sent to reparation centres or to recyclers centre. This network has to be well organized to work appropriately, and its efficiency has to be optimized as much as possible.

How Circular Economy Affects Corporate Strategy

The concept behind Patagonia's business strategy is to manufacture, repair and recycle products, to make them last as long as possible, while creating social and economic value through its activities. In this perspective, circular economy in the fashion industry can be defined as a system whereby everything in the process of making garments is re-used or recycled, including the clothes themselves.

Patagonia's strategy assumes that reselling clothes is crucial for today's fashion brands, considering the amount of products that are already on the market and destined to become waste. However, Patagonia acknowledges the difficulties in implementing such a plan, starting from the creation of adequate infrastructures to support the upcycling stream. It must be noted that Patagonia is no stranger to second-hand, since in 2005 it started collecting garments for recycling via the Common Threads Recycling Program. Then, in 2011, the Common Thread Initiative was born, together with a fruitful partnership with eBay, which represented a significant step for the firm, since it potentially allowed Patagonia to reduce variable manufacturing costs. In fact, through a specific eBay corner, costumers could buy and sell their unwanted Patagonia's items. In 2013, Patagonia launched the Worn Wear project with the purpose to repair and re-use the brand's garments, to extend their useful life as long as possible. In this regard, the company's ultimate goal was to reduce drastically its environmental footprint. When Patagonia receives used products, the first steps is that of "repair" and "re-use", which consists in robust programs to keep people using the clothes that are in good conditions. The programs consist of the re-sale of used clothes through the Worn Wear program (including in-store events, tours and e-commerce platforms), resale of used garments to employees and donations. Also, Patagonia collects and store the items that require small repair and utilize these for teaching skills at Worn Wear events. Lastly, the clothing that is no more treatable is recycled in different ways: upcycled into a new product, mechanically recycled into materials for industrial applications, chemically recycled in closed-loop process (De Giovanni and Zaccour, 2022).

Today, Worn Wear accounts for a considerable part of Patagonia's business and stats show that purchasing a used garment over a new one significantly reduces an individual's apparel carbon footprint by up to a 60% reduction in CO2. The rationale behind Worn Wear is to underline the quality and durability of the brand's items, establishing the value proposition in a particular way, which may be described as counter intuitive. The project displays Patagonia's ability to put the sustainability concept into practice and this is a valuable tool to expand the customer base and strengthen brand loyalty. Therefore, it can be noted that Patagonia's costumers represent a critical resource for the firm. The success of Worn Wear is inspiring other brands to adopt "recommence," as the practice is known. Over the past three years, the second-hand clothing market has grown 21 times faster than the overall apparel industry and is expected to reach $51 billion by 2023.

When analysing corporate decisions, it is important to mention that, since 1985, Patagonia has donated 1% of its sales to protection and restoration of the natural environment. In this perspective, Patagonia devolved cash donations, together with in-kind donations and volunteer time, for more than 89 million dollars to associations working both on the national and international level. Moreover, the company promoted a collaborative initiative with other actors, such as Blue-Ribbon Flies, to encourage other firms to take a step forward in environmental and social sustainability. Thus, in 2002, the "1% for the Plant" organization was born and today it counts more than 1200 members, which are fully devoted to enhance the overall health of the environment and pay very close attention to socio-economic impacts of their corporate's activities. The global network is heterogenous, connected with high impact non-profit partners that align their mission to the Patagonia's one. Brands like Caudalie, Honest Tea and Brushfire Records share the same values as Patagonia and joined the 1% network and, from 2015, induvial donations are an integral part of the initiative. Given this framework, Patagonia can be regarded as a leader in Corporate Social Responsibility and it is, indeed, a benefit corporation, as it signed up for a B certification in January 2012. In order to be regarded as a B Corp, a firm must have an explicit social or environmental mission, just like Patagonia, and a legally binding fiduciary responsibility to take into account the interests of workers, the community and the environment as well as its shareholders. Becoming a B Corp, allowed Patagonia to join high profile, respected companies, gaining a positive reputation on the market and creating an attractive investment opportunity for conscious consumers.

Link Between Circular Economy and Performance

Patagonia has always adopted an environmentally friendly strategy, never implementing direct advertising campaigns on the product but more on the effects their clothes had on nature, even more so in recent times. Furthermore, the fact that there was no direct advertising fuelled a certain hype in the brand. All these factors have translated into excellent economic results for the company, which has always believed in their values and has brought it to a position of leadership compared to other brands that are espousing these themes recently. Thanks to upcycling and its process of circularity, one of the key factors in the creation of products, the procurement of materials, is solved, translates into a reduction in costs and a quick availability of the stock materials (Jalali et al., 2020).

There are various factors that help to give the brand a positive and respectful image which is essential for a company operating in the fashion industry. The elements that can foster a positive impact can be corporate responsibility in and out, cost reduction, ready availability of supplies, footprint reduction, increasing positive brand image and reputation, eco-friendly operations, community center strategies (Maranesi and De Giovanni, 2020). From an economic point of view customers trading in used items receive a fee for the traded in product and Patagonia sells it with a margin after quality curation on its platform. There is a double gain as far as Patagonia is concerned. Firstly, a margin that comes from the repair eventually made or in any case from the process that requires work. On the other hand, in exchange for these products, the company delivers a voucher that can be spent on the website for Patagonia products. The costumer receives this voucher, and his action has a positive impact on the environment, giving a strong sense of social responsibility linked to the brand.

The purpose of the Worn Wear initiative is to increase corporate responsibility, promoting fair labour practices and safe working conditions throughout Patagonia's supply chain. Corporate responsibility is a broad-based movement in business that encourages companies to take responsibility for the impact of their activities on customers, employees, communities and the environment. Companies committed to corporate responsibility also agree to abide by international labour and human rights standards. Patagonia products are produced under safe, fair, legal and human working conditions throughout the supply chain, focusing on a "good supplier list" they have developed in few years and continuously updating (Sacco and De Giovanni, 2019).

From a market perspective, according to ThredUP's, annual resale report, the second-hand market is due to hit $64 billion in the next five years. And by 2024, resale is expected "to overtake the traditional thrift and donation segment. The Worn Wear team goes on the road with repair techs who fix products—regardless of brand—and teach customers how to repair and take care of their gear. In North America, Patagonia's team have been to nearly 120 locations, including stores, specialty retailers, colleges, surf towns, ski resorts and climbing festivals. Reliability, flexibility, finance, and quality are four common aspects that are used to assess economic performance from aspects of product/service quality, production, stock, delivery, supplier and supply chain provided a detailed analysis of flexibility. There are four subareas involved in assessing the effects of practices: supplier, supply, manufacturing, and distribution traceability.

Circular Economy Developments and Business Transformation

The beginning of Worn Wear initiative has to be ascribed in 2005, when it was called "Common Threads initiative": it marks one milestone in Patagonia's history of environmental innovation.

Through the Common Threads Garment Recycling Program, Patagonia started to collect worn-out, old Capilene base layer garments from customers in order to recycle the garments into new polyester (PET) that would have been used to make new filament yarns. The circular economy business was based on a collaboration with TEIJIN, a progressive fabric manufacturer in Japan. Thanks to ECOCIRCLE recycling system, Patagonia's old Capilene garments could technically be processed into new fibers. The main idea was to improve the sustainability practice at Patagonia, as this process had different environmental benefits:

- Given the fact that traditional polyester is made from petroleum, using recycled fibers reduced the fossil fuel-based input needed to produce polyester.
- The improved process also used less energy in manufacturing, reducing also GHG emissions.
- It allowed the waste collection reduction for Patagonia business.

The process was structured as follows: Patagonia set up a reverse-distribution system to collect retired garments from customers and send them to TEIJIN for recycling. Patagonia asked customers to drop garments off in our stores or to ship them directly its Distribution Center in Reno. However, after having conducted some internal analysis and benefits estimation, Patagonia realized that in terms of energy use and CO2 emissions (De Giovanni and Vinzi, 2014), the process of recycling old Capilene garments and shipping them from the U.S. to Japan, was not the option with the least impacts. Indeed, being of on the leading company in sustainability practices, Patagonia was aware of potential gaps and decided to carry on an analysis in which three different scenarios were compared:

A.) Virgin Process: TEIJIN's production of polyester from virgin materials
B.) Locally Recycled Process: TEIJIN's production of polyester using recycled garments that were collected in Japan. (Garments collected at Patagonia Japan locations fit into this scenario).
C.) Recycled Capilene Garment Process: TEIJIN's production of polyester using Patagonia's recycled Capilene garments that were collected in the U.S.

The outputs also revealed that the great impact was represented by the transportation required to move old Capilene garments from customers' closets to Patagonia collection centres domestically. Even if critical, these points could be easily addressed by Patagonia: for the first challenge, the solution was to mail in old Capilene garments rather than making a special trip to drive them to a Patagonia store: this resulted in significant energy and emissions savings.

For the scenario of the second challenge, its finding was considered encouraging as it was an area under Patagonia control. But this was just the beginning. It emerged the supply issue: the fact was that Patagonia products last and so there was not the needed influx of used Capilene base layers. This problem encompassed the second challenge, namely the economic one. Without enough supply, it was not possible to meet the capacity of an industrial recycling machine. The breakpoint came when Teijin moved its recycling facility to China, where there were stricter rules on waste. Patagonia could never align the infrastructure to make it work with the new China-based operation and had to walk away, giving upon the project. The solution was adopted in spring of 2007, when Patagonia expanded the Common Threads Recycling Program. In addition to Capilene garments, it started to accept 100%-cotton T-shirts, Patagonia fleece, as well as Polartec fleece from any brand. Polartec, Inc., is the industry's best-known fleece-fabric manufacturer. Because fleece is made of polyester, Patagonia was able to arrange with TEIJIN to recycle these garments as well into new fiber. By taking back other brands' jackets, the idea was to spark some positive competition, able also to reduce the costs.

Indeed, it took a large capital investment to buy the recycling machines and then more money to keep them running. To this, it should be added also the shipping cost of the old garments, which depends on the distance, the mode of transport, and the cost of fuel. A recent shipment to TEIJIN cost about $1.50 per kilogram, and the processing cost is another $0.50 per kilogram. In addition, buying the recycled fiber usually costs 20% to 30% more than virgin fiber. The provisions relied on the fact that that higher recycling volume and more market competition will make the program more affordable in the future, or that the long-term rise in the price of oil will drive virgin feedstock prices up to a competitive level. So even if the Capilene base layers recycling experiment was not so successful as hoped, it provided some useful learnings: first of all, supply is not an issue. The most important thing was to broaden the spectrum of recyclable items. Secondly, the difficulties related to the infrastructure were solved through new partnerships: the improved network see as collaborator Vertical Knits, a Mexico-based manufacturer that own the entire creation process, from yarn to fabric to garment, and can control scrap collection (Vishkaei et al., 2021). It also started to work with Infinited Fiber, a Finland-based recycling operations that can transform almost any waste—even cardboard—into new fiber. This structure allowed better management of the infrastructure side.

The next step was represented by the evolution of Common Thread Recycling program into Common Thread Initiative in 2011. The idea was to create a partnership with costumers to consume less and reduce its environmental harm.

"This program first asks customers to not buy something if they don't need it. If they do need it, we ask that they buy what will last a long time – and to repair what breaks, reuse or resell whatever they don't wear any more. And, finally, recycle whatever's truly worn out. We are the first company to ask customers to take a formal pledge and be partners in the effort to reduce consumption and keep products out of the landfill or incinerator" declared Yvon Chouinard, Patagonia's founder and owner.

In order to make the process as easy as possible, Patagonia began to collaborate with eBay to launch a new marketplace for costumers, not only to buy but also to sell Patagonia used items. This passage created a new model of consumption: for the first time, a major brand such as Patagonia one actively encouraged its customers to buy and sell used products on eBay. In this way, the firm's slogan "Reduce, repair, reuse and recycle" found application in real terms.

The last circular economy development is adopted when the Common Thread Initiative turned into Worn Wear, in 2013. At the beginning, it made possible for customers to shop for used items in select stores. Then, in 2017 Worn Wear went online and this was the real breakpoint. As already explained, Worn Wear allows customers to buy and sell Patagonia used items. It is essentially Patagonia's hub for keeping a gear in play to cut down on consumption and get more use of stuff clients already own. Since its launch, Worn Wear sustainability outcomes are astonishing: It is estimated that Patagonia has sold 120,000 repurposed items to date (Batten, 2020). Some estimates suggest that buying used garments rather than new ones could reduce an individual's apparel carbon footprint by up to a 60% (Batten, 2020). Furthermore, an increase of 10% in second-hand sales could cut carbon emissions per tonne of clothing by 3% and water use by 4%, if it extends garment life by 50% (WRAP, 2017).

Since its inception, Patagonia has always considered its environmental impact. The entire business model was carried out with sustainability as an absolute principle. Thanks to its continuous investments, today Patagonia is one of the first companies in terms of commitment to the environment and to social issues. From the birth of the Common Thread Recycling Program until the arrival of Worn Wear, the company's circular economy has undergone several developments, in terms of inputs to be processed (at the beginning we only talked about Capilene base layers to get to all the products in cotton) and of actors belonging to the circular economy network (from Teijin partner to Vertical Knits).

These developments allow Patagonia to improve its sustainability performances and enlarge its business, so that today different future actions can be considered: for example, there is the idea to create a first "Rental store" where it will be possible to use some items, without buying them. This experiment could open the doors for new business model in the fashion industry, where the problem of "fast fashion" and its environmental impacts is really challenging our planet.

Business Transformation

To analyse the main business transformation of Worn Wear it is useful to divide this concept into sub-categories, depending on the type of change that has occurred.

Given the fact that business transformation is an umbrella term form making fundamental changes in how a business or organization runs, including personnel, processes and technology changes, for the purpose of our analysis, we can divide it as follows:

Business Process Transformation

Business process transformation means identify those processes that can be improved to reach more efficiency. Patagonia did that when Capilene baselayers experiment did not pan out as expected. Two decisions where taken: enlarge the input of the recycling process, by adding to the Capilene items 100% cotton items and by changing its circular economy network through new actors to recycle its products.

Both actions aimed at maximizing efficiency and cutting the costs: indeed, thanks to enlargement of the input feedstock, Patagonia succeeded in solving the supply-chain scarcity problem on one hand, whereas on the other hand, the identification of new partnerships (such as with Vertical Knits and Infinited Fiber) allowed to solve the infrastructure challenges and save money in terms of logistic expenditure.

Organizational Transformation

Business transformation also implies the redesigning of company's mode of operation. In the case of organizational changes, the focus is in particularly on the employees as they represent a company's most precious resource. The idea at Patagonia headquarters was to close private offices to create open plan working helping to create a culture of communication and collaboration amongst employees. It is worthy to say that, since its beginning Patagonia prioritizes employees,' treatment as demonstrated by the fact that in the 90s it opened an on-site cafeteria for employees serving healthy organic food throughout the day, as part of the company's ethos of taking care of employees as well as customers and the planet.

Cultural Transformation

Even if the company's mindset has been always extremely focused on the respect of the environment and the promotion of social development, through the adaptation of its business itself to such approach, one year after Worn Wear went online, Patagonia announced that its business would have changed: "save our planet" is the mission of the firm since then. Even if from the point of view of concrete commitments, Patagonia had already shown itself to be an environmentally friendly company, with this statement the goal is to strengthen its position and shift the focus of the business from just profit to the business increased focus on their impact.

Digital Transformation

Generally, in this case, what happens is that the business decides to improve its economic performances through the implementation of technologies within its model. The focus is not only on the product but also on the costumer experience: a company needs to rethink its business model with a digital strategic mindset.

When Patagonia decided to partnership first with eBay and then to create its own store (Worn Wear), it relied on innovative technologies that allow the firm to go online and improve its sales of reused items. Costumers' experience was central: at the beginning the idea was to provide used items only in selected stores, but in this way most of the costumers could not join such initiative and so the solution was to create an online platform in which, regardless of where you are in the world, you can buy and sell recycled products.

Figure 3. The most relevant programs impacting on Circular Economy

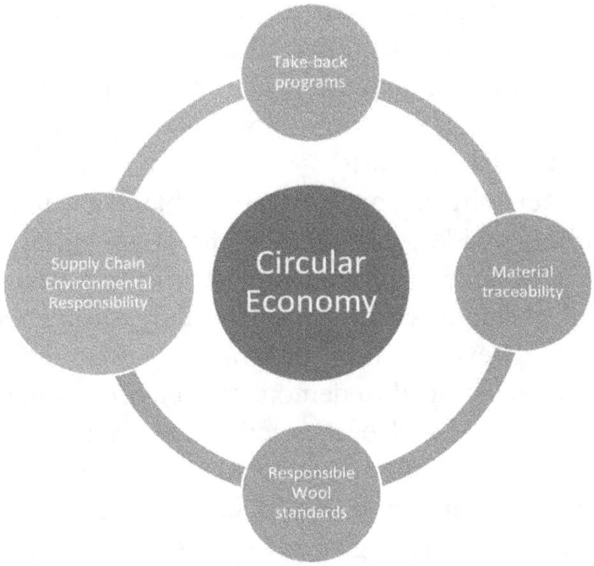

ENVIRONMENTAL RESPONSIBILITY

To reinforce the circular economy strategy, Patagonia has initiated several environmental programs. Those contribute to the circular economy strategy are displayed in Figure 3.

- **Take back program**. Most of the clothing industry is based on a take-make-waste model, which fails to take responsibility for what happens to clothing once someone no longer needs it. Without a solution in place, global textile waste is expected to increase 60 percent annually until 2030, when it's estimated to hit 148 million tons per year. And the issue isn't just volume, but what happens once those textiles are sent to landfills. Landfills release greenhouse gases, impact wildlife habitats and pose a risk to air and water quality—all of which disproportionately affect minority and low-income communities who often live near these sites. To help build a circular clothing system—one that prioritises a more efficient use of resources and upcycles waste into new clothing so it can be reused multiple times—Patagonia launched the Take-Back Program, repurposing your old Patagonia tees into new ones.
- **Supply Chain Environmental Responsibility**. The purpose of Patagonia's Supply Chain Environmental Responsibility Program

is to measure, reduce and eliminate the environmental impacts of manufacturing Patagonia products and materials. Patagonia implements the program at supplier facilities all over the world and cover a broad range of impact areas, including environmental management systems, chemicals, water use, water emissions, energy use, greenhouse gases, other air emissions and waste.

- **Material traceability**. This program is based on internal standards, utilizes industry-wide tools, such as the Higg Index, and recognizes third-party certification programs, such as the bluesign® system, as ways that the suppliers can show how they are meeting Patagonia's expectations. Beyond the program's minimum requirements, suppliers are encouraged to demonstrate environmental excellence by implementing better and best practices, so Patagonia can recognize them as environmentally responsible supply-chain partners. The supply chain is what textile and other manufacturers use to describe everything from the crops grown to make yarn and the sewing of the fabric into garments, to shipping finished clothes to warehouses, stores and the customers' front porches. Supply chains are long and complex, involving many entities between the origin of a fiber and the finished product. Patagonia relies on the work of thousands of factories, mills, spinners, processors, recyclers and farms to make the products.
- **Responsible Wool Standards**. The majority of products are made with the most environmentally sustainable materials available, including organic cotton, Traceable Down, wool that meets the Responsible Wool Standard, natural rubber and various recycled materials. The company strives to establish traceability of the materials by mapping out all of the entities in these supply chains, and also implementing strong chain-of-custody guidelines for suppliers, to reduce the possibility that less sustainable alternatives are swapped or mixed with more sustainable materials. Wool is a natural fiber that insulates, breathes and lasts for a long time. Producing wool, however, is a resource-intensive job. It requires vast amounts of land for grazing sheep, water to clean the fiber, chemicals to treat the wool and dyes to color the finished product. In addition, not all wool supply chains provide humane treatment of animals, which is why Patagonia relies on the Responsible Wool Standard to source the wool in the products.

These programs are very urgent in the textile industry due to considerable impacts that it has on the environment, which is summarized in Figure 4.

Competitor's Analysis: The North Face and Save the Duck.

In terms of competitors, it is useful to compare Patagonia business approach and more specifically Worn Wear initiative to see where lies the unique value of our business case.

A valuable competitor is represented by The North Face, that is an outdoor company in the market, but it focuses more on fashion. It was founded in 1968 in San Francisco, CA. The company's initial offerings were for hardcore climbing enthusiasts since they had almost every kind of mountaineering gear in their repertoire. However, sometime in the late '90s, and most notably after the VF Corporation acquisition, they started dealing in rugged yet elite street fashion. Today, The North Face, headquartered in Denver, CO, has more than 5000 stores in over 50 different countries. They still stay true to their original climbing gear designs but have shifted their focus more toward fashion.

In terms of sustainability, North Face is committed in reducing its environmental impact by promoting the production founded on sustainable materials and by helping customers to buy sustainable. The main ideas are indeed to preserve the items as more as possible and to recycle them to give products a second life. Similar approaches between the two firms can be founded, but the difference is that Patagonia is really acting right now in this direction. Indeed, even if the North Face has different programs designed to combat waste and extend the life of their products, only in 2022 the first products specifically designed to support circularity will be launched.

The most similar initiative to Worn Wear is represented by Clothes the Loop of The North Face. Even in this case, the idea is to regenerate the life of used items (regardless of whether they are North Face designer clothes or not) through a process in which the customers play a central role: they deliver clothing and footwear they no longer wish to wear at The North Face stores. It doesn't matter what condition they are in or what brand they are. The items collected in the containers are sent to a recycling centre, where they are carefully divided into 400 different categories. They are then reallocated for reuse, so they can extend their life cycle, or recycled into raw materials for use in insulation products, carpet and toy fillings, and textile fibers for new clothing. The main difference here is that whereas Worn Wear went online already in 2017, Clothes the Loop initiative remains a programme available only on physical store: this feature is overwhelming discriminant in terms of audience engagement and expected outcomes. The fact is that, even if the two programmes were launched in the same year, Patagonia immediately

Figure 4. The environmental impact of the textile production by EchoChain (2022)

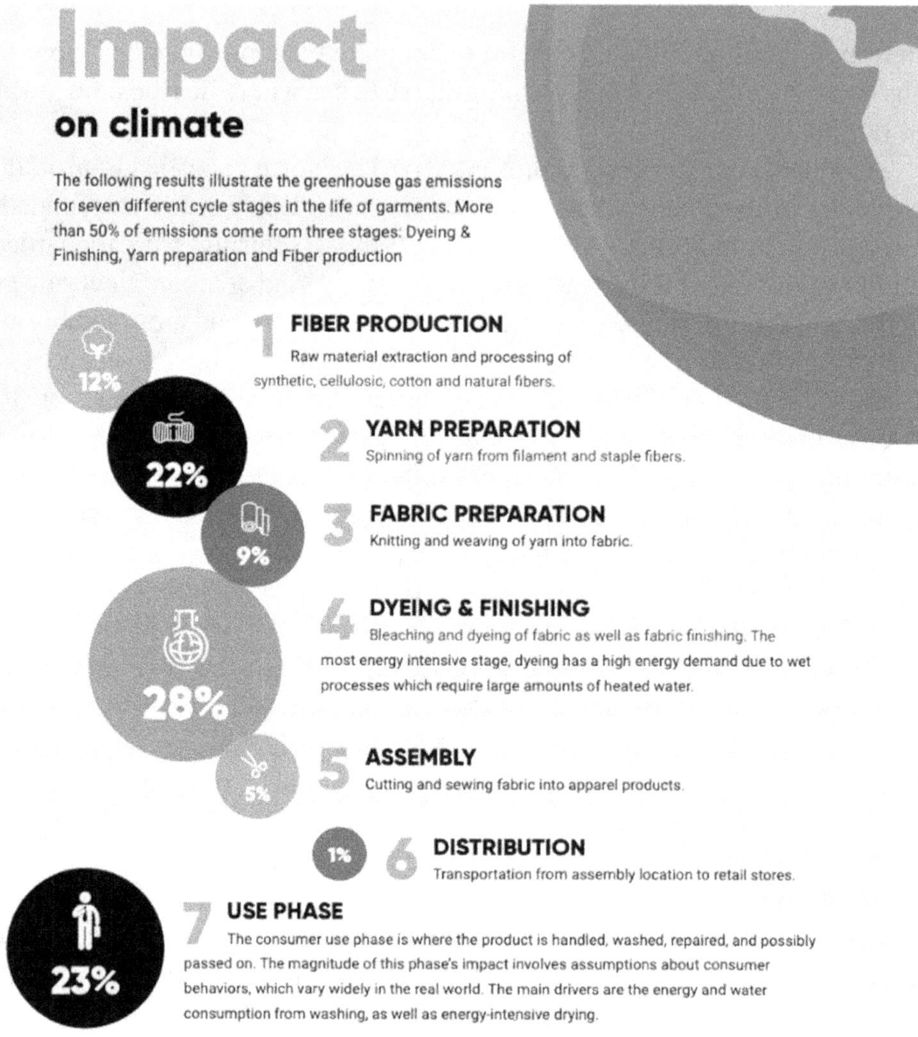

had the insight that limiting Worn Wear initiative to the stores would have impacted its effectiveness: in order for the share of the used clothing market to have a significant impact in environmental terms it was necessary to expand participation as much as possible and this is possible through an online platform, such as that of Worn Wear., as demonstrated by the fact that Worn Wear still only accounts for $5 million of Patagonia business. New products still reign supreme, and customers need more of an incentive to trade in their used goods (De Giovanni, 2022).

As usually, the unique value of Patagonia is its foresight meaning the intrinsic ethos of the company itself that has historically pushed it to do better for the sustainability of its performances.

Another interesting comparison can be made with the younger Italian *Save the duck*. This brand was launched in 2012 by Nicolas Bargi, nephew of the *Forest* founder Foresto Bargi, with the aim to create a product that cares about the animal, the environment and the society. Following the new ethical paradigm that also distinguishes Patagonia, *Save the duck* has done many collaborations and initiatives to raise people's awareness about sustainability. However, we can find a clear difference between the two companies that is related to the use of products with animal origin such as the duck feathers. Indeed, even if with sustainable practices, Patagonia uses products with animal origin. Reading its Traceable Down standard, we can clearly see that they assure to the customers that their products have been made without causing unnecessary harm to animals, in compliance with local legal animal welfare regulations and in collaboration with farms that do not engage in force feeding or live plucking. On the contrary, *Save the duck* is born to give an alternative to the animal sources.

POSSIBLE IMPLICATIONS IN TERMS OF SMART CITIES

To analyse the Worn Wear start-up based on its linkage with the development of a smart city, it comes as a form of weakness because the start-up has limited or no connection with smart city applications. There are lots of innovative and sustainable technologies emerging for smart textile now which Worn Wear Patagonia can learn from. As a form of recommendation, to suggest new ways to expand their business, it can learn from some of these smart textile solutions mentioned below to better expand its business standards, solutions, and practices.

Smart Bins Platform

The Zero Panik Textile Recycling Company has direct access to the textile waste market and actively promote the brand using smart bins. The bins are for collections of old textiles or textiles to be thrown away, then, the bins are built with voice guidance, and AI tech that allows for instant notification when bins are full. The functions of the smart bin include rewarding contributors for their donations, incentivizing more donors. It also tracks recycling efforts, and when it is full this smart bin will lock down and notify the drivers for pick-up. For drop off, the on-screen notice will alert contributors of the time of scheduled pickup and refer them to the next nearest Zero Panik Smart Bin Location.

Ekocharita company in Slovakia collects, pre-sorts, and sells used clothing, footwear, bed sheets, linen in wholesale, either for further reuse or recycling. The company uses Sensoneo cloud-based platform that enables monitoring of wastebins, and optimization of waste collection routes, frequencies, and vehicle loads. The Sensoneo technology enables the company to digitize their bin infrastructure and combine fleet and driver schedules with daily demands for collection of textile containers from customer and citizens. With Sensoneo, Ekocharita also receives daily many requests through different channels (Facebook, email, phone.) from cities and citizens.

Recycling Smart Clothes

An example of smart clothes is The Hug Shirt, developed by cute circuit_that lets you send hugs over distance, with the presence of actuators/sensors that recreate the sensation of touch and the emotion of a hug to the Hug Shirt of a distant loved one. It has a Sustainable zero-waste design and manufacturing made with more of a simple sensor simply compatible with Hug Shirt App. The textile has a Standard for Sustainability, making it possible for recycling. The sensor in the shirt is made in such a way that they have excellent long-term sustainability benefits over their entire product life cycle. The hug shirt is a shirt that recreates the sensation of a human hug. So, if you are far away from someone, you can use the app to send a hug to someone who is wearing a hug shirt by downloading and using the app. According to the website, the smart textiles within the shirt will squeeze you as good as any friend would. This product seems increasingly relevant in the time of the pandemic and social distancing.

Since anything you buy from Patagonia that is no longer reusable or repairable in any way can be returned so that it can be recycled into new fabrics, Worn Wear can also use this opportunity to make these fabrics into smart clothes.

CONCLUSION

Patagonia is a company that has successfully implemented a fascinating business model that illustrates the importance of expanding their profit margins, while promoting an environmental green manufacturing process. In the clothing industry today, Patagonia is a revered corporation since they have an effective and distinctively unique concept that prioritizes the sustainability of the planet over profit. This initiative was started in 1985 by an individual Yvon Chouinard a passionate mountain climber who had a distinct love for our planet. He used his love for the environment to create one of the first environmentally friendly companies that has flourished in a capitalistic industry.

In 2013 the company decided to create a project that was designed with a pure purpose of extending the life of their manufactured products. The "Worn Wear Project," was incredibly unique because it created a platform that allowed the consumers to take advantage of incentive programs. The incentive was to give the consumers a reusable option when it came to buying more products, in other words the company would offer customers gift vouchers towards their next Patagonia purchase if they recycled the old products. This project was correlated directly with the companies' main values; producing the best quality products while causing limited harm to the environment during the manufacturing process, and lastly utilizing their profits to sustain and protect the environment.

In the fashion industry today, a big number of products is destined to become waste. Therefore, in 2005 Patagonia initiated a program that would have a significant effect on the recurring problem of waste. The company began to formulate partnerships via the Common Threads Recycling Program where they were able to start collecting secondhand garments, this initiative was the starting block for the Worn Wear Project. Thanks to this initiative now today the Worn Wear project accounts for a considerable amount of the companies' numerical statistics signifying that the recycling of used garments significantly reduces the carbon footprint up to 60 percent.

The Worn Wear Project has been used as a benchmark tool by a considerable number of firms trying to incorporate a more sustainable approach during the manufacturing process. The Xero Panik Textile Recycling Company incorporated a likewise approach utilizing the concept of incentivizing consumers just like the Worn Wear project created by Patagonia. The company wanted to create a system that would significantly reduce the amount of textile waste by implementing smart bins in specific locations within cities. Smart bins allow the citizens to have easy and reliable access to dispose of their textile waste, while simultaneously creating a reward system for the citizens, incentivizing more donors, and reducing overall waste.

Patagonia and the founder Chouinard created a revolutionary concept that incorporated the idea sustainable manufacturing of products over importance of profit. His business model is being studied and utilized nowadays by several firms in this sector, but many of them still struggle to understand and implement the concept of circular economy in their business model. In the words of Chouinard, "Purpose is not a strategy," meaning that firms cannot change their mission through marketing strategies.

REFERENCES

Caniato, F., Caridi, M., Crippa, L., & Moretto, A. (2012). Environmental sustainability in fashion supply chains: An exploratory case based research. *International Journal of Production Economics, 135*(2), 659–670. doi:10.1016/j.ijpe.2011.06.001

De Giovanni, P. (2016). Coordination in a distribution channel with decisions on the nature of incentives and share-dependency on pricing. *The Journal of the Operational Research Society, 67*(8), 1034–1049. doi:10.1057/jors.2015.118

De Giovanni, P. (2019). When feature-based production capabilities challenge operations: Drivers, moderators, and performance. *International Journal of Operations & Production Management, 40*(2), 221–242. doi:10.1108/IJOPM-04-2019-0309

De Giovanni, P. (2020). *Dynamic quality models and games in digital supply chains*. Springer Nature.

De Giovanni, P. (2022). Leveraging the circular economy with a closed-loop supply chain and a reverse omnichannel using blockchain technology and incentives. *International Journal of Operations & Production Management, 42*(7), 959–994. doi:10.1108/IJOPM-07-2021-0445

De Giovanni, P., & Folgiero, P. (2022). Strategies for the Circular Economy. Circular Districts and Networks. Giappichelli-Routledge. ISBN: 9788892123876.

De Giovanni, P., & Vinzi, V. E. (2014). The benefits of a monitoring strategy for firms subject to the Emissions Trading System. *Transportation Research Part D, Transport and Environment*, *33*, 220–233. doi:10.1016/j.trd.2014.06.008

De Giovanni, P., & Zaccour, G. (2022). A selective survey of game-theoretic models of closed-loop supply chains. *Annals of Operations Research*, *314*(1), 1–40. doi:10.100710479-021-04483-5

Del Giudice, M., Chierici, R., Mazzucchelli, A., & Fiano, F. (2020). Supply chain management in the era of circular economy: The moderating effect of big data. *International Journal of Logistics Management*, *32*(2), 337–356. doi:10.1108/IJLM-03-2020-0119

Duhaylongsod, J. B., & De Giovanni, P. (2018). The impact of innovation strategies on the relationship between supplier integration and operational performance. *International Journal of Physical Distribution & Logistics Management*, *49*(2), 156–177. doi:10.1108/IJPDLM-09-2017-0269

EchoChain. (2022). *The data behind Patagonia's no-logo decisions (and why it might be wrong).* Available at: www.echochain.com

Farooque, M., Zhang, A., Thürer, M., Qu, T., & Huisingh, D. (2019). Circular supply chain management: A definition and structured literature review. *Journal of Cleaner Production*, *228*, 882–900. doi:10.1016/j.jclepro.2019.04.303

Genovese, A., Acquaye, A. A., Figueroa, A., & Koh, S. L. (2017). Sustainable supply chain management and the transition towards a circular economy: Evidence and some applications. *Omega*, *66*, 344–357. doi:10.1016/j.omega.2015.05.015

Guide, V. D. R. Jr, & Van Wassenhove, L. N. (2006). Closed-loop supply chains: An introduction to the feature issue (part 1). *Production and Operations Management*, *15*(3), 345–350. doi:10.1111/j.1937-5956.2006.tb00249.x

Hofmann, F., & Jaeger-Erben, M. (2020). Organizational transition management of circular business model innovations. *Business Strategy and the Environment*, *29*(6), 2770–2788. doi:10.1002/bse.2542

Jalali, H., Ansaripoor, A. H., & De Giovanni, P. (2020). Closed-loop supply chains with complementary products. *International Journal of Production Economics*, *229*, 107757. doi:10.1016/j.ijpe.2020.107757

Köhler, J., Sönnichsen, S. D., & Beske-Jansen, P. (2022). Towards a collaboration framework for circular economy: The role of dynamic capabilities and open innovation. *Business Strategy and the Environment*, *2*, 1–30. doi:10.1002/bse.3000

Köksal, D., Strähle, J., Müller, M., & Freise, M. (2017). Social sustainable supply chain management in the textile and apparel industry—A literature review. *Sustainability*, *9*(1), 100. doi:10.3390u9010100

Maranesi, C., & De Giovanni, P. (2020). Modern circular economy: Corporate strategy, supply chain, and industrial symbiosis. *Sustainability*, *12*(22), 9383. doi:10.3390u12229383

Sacco, A., & De Giovanni, P. (2019). Channel coordination with a manufacturer controlling the price and the effect of competition. *Journal of Business Research*, *96*, 97–114. doi:10.1016/j.jbusres.2018.09.001

Saeed, M. A., Waseek, I., & Kersten, W. (2017). Literature review of drivers of sustainable supply chain management. *Proceedings of the Hamburg international Conference of Logistics (HICL)*, 159-184.

Seuring, S., & Müller, M. (2008). From a literature review to a conceptual framework for sustainable supply chain management. *Journal of Cleaner Production*, *16*(15), 1699–1710. doi:10.1016/j.jclepro.2008.04.020

Shatri, A. (2021). *Complete SWOT Analysis of Patagonia. One of the most sustainable and eco-friendly retailers.* The Digital School. Available at: https://iide.co/case-studies/swot-analysis-of-patagonia/

Srivastava, S. K. (2007). Green supply-chain management: A state-of-the-art literature review. *International Journal of Management Reviews*, *9*(1), 53–80. doi:10.1111/j.1468-2370.2007.00202.x

Tseng, M. L., Tan, R. R., Chiu, A. S., Chien, C. F., & Kuo, T. C. (2018). Circular economy meets industry 4.0: Can big data drive industrial symbiosis? *Resources, Conservation and Recycling*, *131*, 146–147. doi:10.1016/j.resconrec.2017.12.028

Vishkaei, B. M., Fathi, M., Khakifirooz, M., & De Giovanni, P. (2021). Bi-objective optimization for customers' satisfaction improvement in a Public Bicycle Sharing System. *Computers & Industrial Engineering*, *161*, 107587. doi:10.1016/j.cie.2021.107587

Walter, G., Knizek, C., von Koeller, E., O'Brien, C., Hardin, E. M., Young, D., & Cordes, F. (2020). *Your supply chain needs a sustainability strategy*. Academic Press.

Zsidisin, G. A., & Siferd, S. P. (2001). Environmental purchasing: A framework for theory development. *European Journal of Purchasing & Supply Management*, *7*(1), 61–73. doi:10.1016/S0969-7012(00)00007-1

Chapter 10
The Waste Cascade in Dell Reconnect With a Focus on Plastic Packaging

Pietro De Giovanni
https://orcid.org/0000-0002-1247-4807
Luiss University, Italy & Luiss-X.ITE Research Center, Italy

EXECUTIVE SUMMARY

The Dell Reconnect program is a clear example of a how the circular economy aims at social development and inclusion through a tripartite of strategies consisting of reuse, refurbish, and recycle. Hence, this chapter explores the peculiarities of the circular economy implemented by Dell Reconnect, as well as all unexploited areas, such as the adoption of public funds and the use of digital technologies for smart circular economy systems.

THE DELL RECONNECT PROGRAM

The Dell Reconnect project is a partnership between Dell and Goodwill agencies spread across the US, which target the recycling of electronics (Maranesi and De Giovanni, 2020). The US consumers can drop off all non-used electronics at Goodwill location. See Ramani and De Giovanni (2017), Jalali et al. (2022), and De Giovanni and Ramani (2017) for a full description of the Dell Reconnect case.

The workers analyze and evaluate the electronics and then decide what to do with it: resell it as is, refurbish, resell, or recycle. Consumers receive

DOI: 10.4018/978-1-6684-5001-7.ch010

a rebate proportional to the electronics' conditions (De Giovanni, 2022). Goodwill makes the environment cleaner and creates green jobs, which implies an overall analysis of the social and environmental implications of Dell Reconnect and all complementary aspects of the business model (Jalali et al., 2020).

Indeed, Dell and the Goodwill Agencies require to identify coordination mechanisms to reach common goals. Several mechanisms have been suggested in the literature to manage such types of relationships (Sacco and De Giovanni, 2019; De Giovanni, 2016). According to the territory to reach, the network of relationship to close the loop can be organized in several ways (for an overview, see De Giovanni and Zaccour, 2022). In fact, as demonstrate in Vishkaei et al. (2021), the optimization of a network that is supposed to manage both forward and reverse activities can be very difficult to optimize, especially when several types of performance indicators are involved (De Giovanni and Vinzi, 2014).

Dell Reconnect and the Waste Cascade

When each item arrives at a Goodwill location, it is checked and positioned inside the waste cascade. The returned items can either be reused, refurbished, or recycled. The items can be classified as *Reuse,* when the electronics are in good conditions and ready to be reused. Instead, the items can be classified as *Refurbish,* when they require an upgrade. If electronics belong to these clusters, they can be resold in a Goodwill location. Otherwise, they are classified for *recycling* activities. A summary of these options is displayed in Figure 1.

By implementing this circular economy strategy, Dell Reconnect enables avoided 1 billion pounds of e-waste every year from the landfill through 2019, which is equivalent to 2,000 jumbo jets, while fixing high-level targets from 2020 onward. Beside the environmental benefits, the circular economy system contributes to the purpose of Goodwill's agency goal, which is putting people back to work. Indeed, the quality of returns makes the difference since their classification is highly linked to the return's residual value and, furthermore, the possible connection with the brand value (De Giovanni, 2020). According to De Giovanni (2021), the return residual value can be checked and monitored using digital technologies, to ensure a systematic verification of the items' conditions (De Giovanni, 2021).

Figure 1. DellReconnect's opportunities

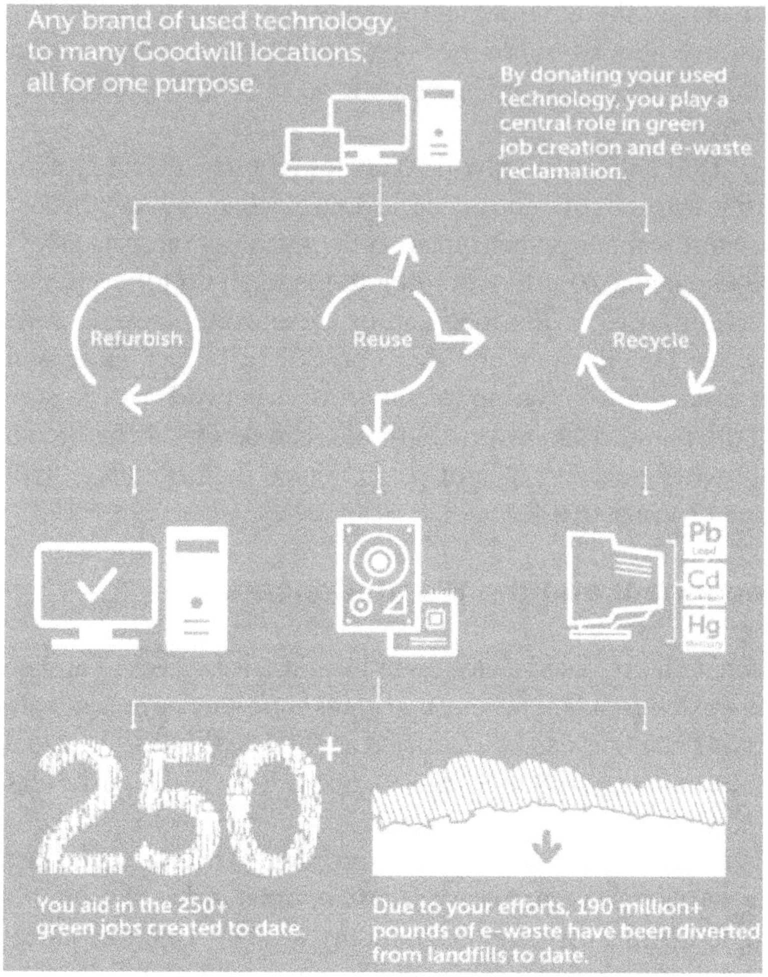

HOW TO ADDRESS THE PROBLEM OF PLASTIC IN DELLRECONNECT

While DellReconnect focuses on the circular economy of electronics, this chapter analyzes the possible implications of this framework in the circular economy of plastic. In fact, the circular economy of electronics includes lots of plastics, which is not treated properly in most of the cases. Figure 3, which has been developed by Statista.com, shows that the plastic is seldom treat for recycling purposes, even in the electronics sector. Consequently, circular

Figure 2. Waste cascade for Dell Reconnect

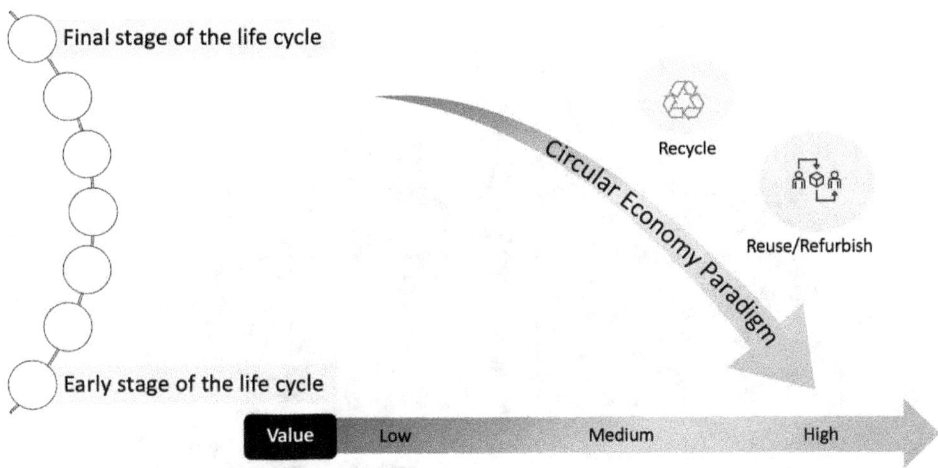

economy systems must be activated to guarantee that the plastic is treated properly, that is, avoiding landfill, compost, and energy recovery.

Plastic is inevitably part of our daily life. Plastics are among the first products made entirely by humans and not found in nature, even though they are organic substances. Indeed, plastics are derived from raw materials including oil, natural gas, coal, common salt, and other natural products. Figure 4 displays the evolution of the plastic treatments in Europe, made by Statistica.com. Accordingly, the recycling activities have been improved over time but they still represent less than 30% of the total amount recycled.

The sectors and the applications of this material are various and diversified, thanks to its chemical and mechanical characteristics. In fact, the advantageous characteristics of plastics compared to metallic and non-metallic materials are different: ease to process, cost efficiency, ability to colour for future usages, use it for removing noises, thermal applications, electrical and mechanical insulation, avoid corrosion, chemical applications, water repellence and the ability to keep the original characteristics even in case of mould, fungi and bacteria.

Since the 1950s, global plastics production has grown at an average rate of 8,7%, with only two downs during the oil crisis in 1970s and financial crisis in 2008. Overall global plastics production has grown nearly 9 times since the 1970s compared to 4.5 times for aluminium and 2.5 for steel and it has decreased for the first time in 2020, shifting from 368 Mt in 2019 to 367 Mt in 2020 (Ambrosetti 2013). Consequently, firms, governments, and institutions,

Figure 3. How the plastic is treated by the end of the life cycle

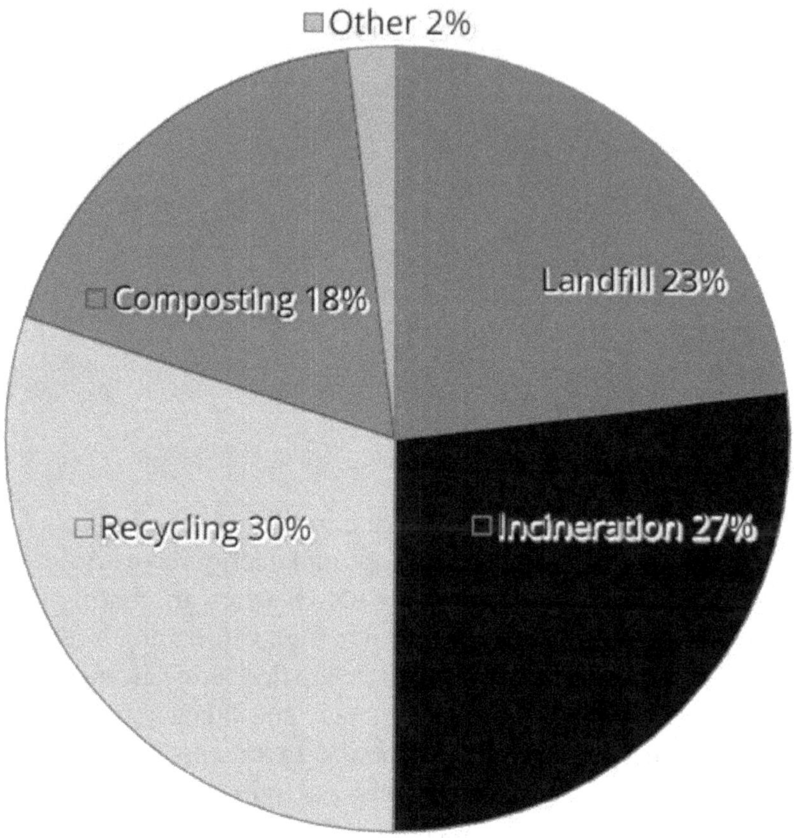

Figure 4. Changes in the plastic treatments over time

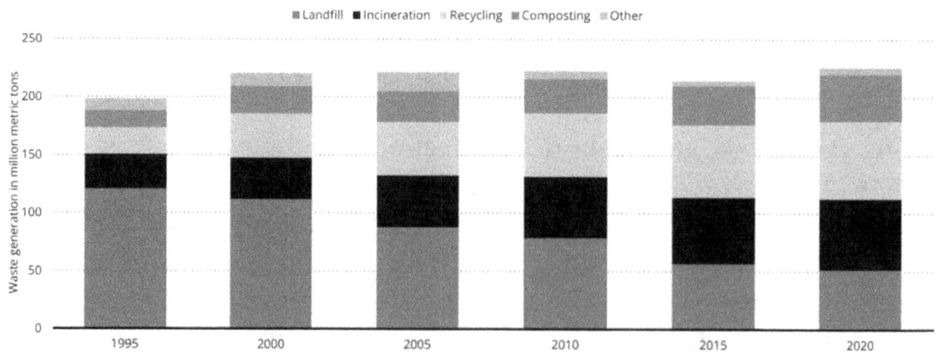

including circular supply chain as the Dell Reconnect, should make an effort to reduce the plastic as much as possible. First, because plastics products are present in many different sectors: from automotive to agriculture, from the construction industry to transports, from the food and beverage to health industries. Second, because plastic can be composed of different polymers combined with chemical additives, it can be easily turned into many different applications, forms, and solutions (EEA 2021). The most commonly used polymers are mainly derived from a few basic petrochemicals, including: ethylene, propylene, butadiene, benzene, toluene, meta and para xylene and styrene. Within the most used and know types of plastics there are: the polyethylene terephthalate (PET), mainly used for beverage bottles, medicine jars, clothing and carpet fibre; the high-density polyethylene (HDPE), mainly used for containers for milk, motor oil, shampoos and detergents; lowdensity polyethylene (LDPE), mainly used for sandwich bags, squeezable bottles and plastic grocery bags; polypropylene (PP), mainly used for yoghourt pots, prescription bottles, syrup bottles and films; polyvinyl chloride (PVC), used for all kinds of pipes and tiles (EEA, 2021). Due to its peculiarities, on one side plastics market is characterised by a high level of innovation and substitution of materials. On the other hand, the plastics market during the last decades showed how some typologies of plastics are more suitable for certain applications only. Considering all the plastics end-use markets in the EU, packaging represents the largest one with more than 40% of the total and about 60% of post-consumer plastic waste. In the packaging industry, the most used types of plastics are polyethylene, polypropylene and PET.

The flexible packaging is mainly made of polypropylene, which represents about 30% of all plastic material in the world thanks to its countless applications and, due to its technical properties. Polypropylene is 100% recyclable and it lends itself to countless uses, from packing cigarette packs on high-speed machines to food packaging, to labels, thanks to its transparency and heat-shrink characteristics that allow it to meet high quality standards by adapting to different needs. The barrier properties against atmospheric elements allow its application for the food sector, guaranteeing the safety of products and extending the shelf-life, i.e. postponing the expiry date of packaged foods. One very popular application is the Bioriented Polypropylene (BOPP). To give some technical details, it means that the plastic film is pulled in two different directions during production, in order to improve its strength, flexibility, and brightness. BOPP is made by extruding the plastic through a circular die followed by expansion cooling (Maddah 2016).

Figure 5. How plastics harms the environment

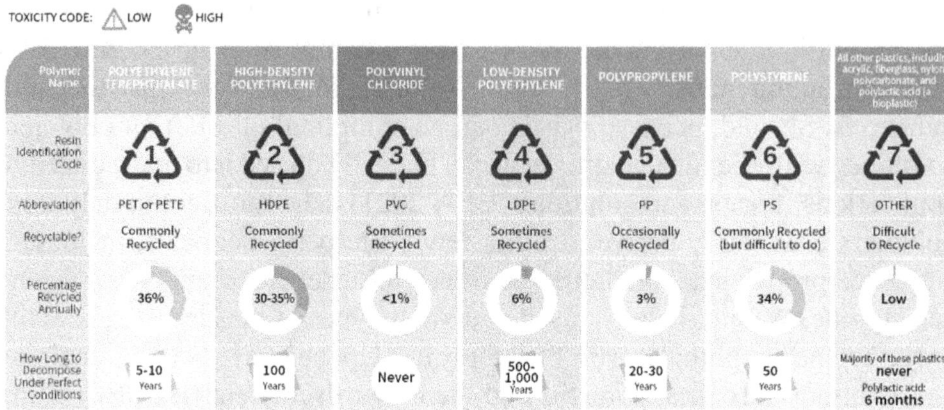

This kind of film provides resistance to moisture vapour to keep snacks crisp and fresh tasting and provides a heat-sealable layer (Maddah 2016). It is mostly used in packaging of bakery products, adhesive tapes and snack food packaging. The BOPP physical, chemical and mechanical characteristics, plus its relatively cheap price, have made this material the most used for flexible packaging in the food and beverage industry application. 90% of all plastics are derived from virgin fossil feedstocks and account for about 6% of global oil consumption which is equivalent to the global aviation sector (Matthews et al., 2020). In fact, plastic material has been used and produced for decades only from a profit perspective, without considering its impact on the environment. However, BOPP is a fossil-based material and, incentivised by the new sustainable regulatory framework, the private sector is trying to produce market feasible solutions to respect the new environmental requirements. Indeed, several non-fossil alternatives and strategies have come out in recent years. From the design of the packaging to the elimination of single use plastic and the substitution of virgin plastics with other feedstocks. Although theoretically the transition towards a plastic free world seems easy and obvious to the public opinion, the reality shows a more complex scenario. Replacing and rethinking plastics is a hard task for the private sector.

The Regulation for Plastic: An Example of Plastic Packaging

During recent years, a new sustainable regulatory framework that sets the rules and the guidelines in the plastic market in Europe has been developed and it is still in development. Indeed, plastic legislations and normative trends have evolved embracing social and environmental values, with the aim of decreasing the negative impacts of the plastic world on society and the environment. In December 2015, the Commission adopted an EU Action Plan for a circular economy. There, it identified plastics as a key priority and committed itself to 'prepare a strategy addressing the challenges posed by plastics throughout the value chain and taking into account their entire lifecycle'. The challenge that firms and circular supply chain like Dell Reconnect face is to accommodate the collection and the recycling of plastic into the whole business model. While Goodwill agencies cannot do much with recycled plastic, Dell can still use it for some of their goods.

In 2017, the Commission confirmed it would focus on plastics production and its use, working towards the goal of ensuring that all plastic is recyclable by 2030 (EU Commission 2018). In 2018 the Commission presented the European Strategy for Plastics in a Circular Economy that shows the key commitments for an integrated action. The Strategy wants to build a new plastics economy, where the design and production of plastics and consequentially of the plastic products fully respect the three R's rule: reuse, repair and recycling (EU Commission 2018). According to the Strategy, the actors involved in the plastics industry should: improve design and support innovation to realise plastic products easier to recycle; expand and promote the separate collection of plastic waste, to ensure quality inputs to the recycling industry; expand and modernise the EU's sorting and recycling capacity; create viable markets for recycled and renewable plastics. While setting the targets and the strategy to follow, the Commission stresses the importance of Extended Producers Responsibility (EPR) to reach the objectives. According to the Organisation for Economic Co-operation and Development (OECD), EPR can be defined as an environmental policy approach in which a producer's responsibility for a product is extended to the post-consumer stage of a product's life cycle. EPR legislation, in principle, shifts the responsibility for, and costs of, negative environmental externalities of products from tax payers to producers, consistent with the "polluter pays" principle. EPR is intended to achieve environmental improvements throughout the product life cycle and has two primary environmental goals. The first is to provide incentives for manufacturers to design resource efficient and low impact products (e.g., ecodesign). The second is to ensure effective end-of-life collection, environmentally-sound treatment of collected products and improved reuse and recycling.

EPR principle was already indirectly invoked by other regulations, like the Packaging and Packaging Waste Directive (94/62/CE) in Article 7, that requires the Member States to take necessary measures to ensure that systems are set up for the collection and recycling of packaging waste. Then, it has been explicitly mentioned in the Single Use Plastic Directive (2019/904) in Article 8, that states that Member States shall ensure that extended producer responsibility schemes are established for a precise list of products.

The development of EPR in Europe has contributed to improvements in waste prevention, reuse and recycling. Related to plastic, in the EU it is possible to summarise the EPR schemes in three macro categories: 1) product take back requirements, which commonly involve establishing either mandatory or voluntary recycling and collection targets for specific products or materials, and assigning responsibility to producers or retailers for end-of-life management to achieve these targets. Applied to our case, Dell can use the Dell Reconnect project to improve the EPR and achieve higher environmental targets. 2) advanced disposal fees (ADF), which are fees levied on individual products at the point of purchase, based on estimated costs of collection and treatment. This will represent a less attractive alternative for Dell Reconnect that will need to penalize all Goodwill agencies that are not respecting the agreements. In fact, the fees may be used to finance end-of-life management of the products. 3) Deposit Refund Systems (DRS), which adds a surcharge on individual products at the point of purchase. The entire fee, or a portion of it, is refundable when the used product is returned to the point of sale or at specified waste management sites. The aim is to encourage take-back of the used product rather than to cover costs. Indeed, this measure applied to the Dell Reconnect project will lead to a more coordinated and collaborative environment between Dell and the Goodwill agencies.

To evaluate EPR role and effectiveness, an article published in 2020 by four scientists of the Technical University of Denmark analyses 40 management scenarios focused on plastic packaging waste generated by Italian households were investigated, and their environmental performance (via a consequential life cycle assessment) and the economic sustainability of their waste value chain (via a cost-benefit analysis for each stakeholder) were compared to the recycling targets. The research states that extended producer responsibility policies (EPR) are crucial to reach environmental and economic sustainability because they are the connection link between the different stakeholders and can influence all of them. Furthermore, EPR plays an important role in assessing the requirements for recycling plastics, which is the fundamental step in implementing a circular strategy for this material, and observing food safety requirements, that is a conditio sine qua non for the Food Contact Material (FCM). On the one hand, plastic packaging reduces food waste by extending the shelf-life (Ellen MacArthur Foundation et al., 2016; White and Lockyer 2020) and prevents spoilage from microorganisms that are one of the main causes of food attack from microorganisms. On the other hand, as stated by Matthews et al. (2020), the majority of plastic used for packaging is single use, with 95% of material value lost to the economy each year with a value of up to USD 120 billion. Furthermore, Matthews et al. (2020) highlight plastic recycling limitations when it comes to using recycled plastic for FCM. As demonstrated also by Geueke et al. (2018), different groups of contaminants, oligomers, additives and their degradation products, as well as chemicals derived from previous (mis)uses, have regularly been reported in recycled plastic. Indeed, virgin and recycled plastic packaging have to observe respectively the EU food contact material (FCM) regulations and the Regulation 282/2008 on recycled plastic materials and articles intended to come into contact with food. The goal of these regulations is to lay down the criteria that ensure plastic packaging and recycled plastic packaging can be safely used for food contact. The impact of plastic packaging is often misinterpreted by consumers. Otto et al. (2020) demonstrate how the consumer perception differs highly from the scientific conclusion of environmentally friendly sustainable packaging. According to their research, "the environmental impact of paper/cardboard and metal are rated in line with the scientific measure by consumers, whereas plastic packaging is underestimated and glass and biodegradable plastic packaging are highly overestimated". Indeed, the tendency highlighted shows that for most consumers plastic represents a low-quality product, that is associated with negative perceptions and feelings (Fernqvist et al., 2015), whereas for the scientific facts plastic represents overall the most sustainable packaging choice (Otto et al., 2020) compared to metal, glass, and paper. It is common for consumers to think about the packaging itself but not about its function and its importance for food. Many LCA's have been conducted for food packaging, but "few consider the interaction between the packaging and packaged food, although it is widely acknowledged that this interaction plays a key role for the environmental performance of food packaging". Furthermore, another study conducted by Kan and Miller (2022) shows the plastic packaging environmental responsibility in food products, considering its climate change and energy use impacts. The study carries out an analysis of 28 studies that conduct an LCA of food products to quantify the impact of plastic packaging relative to the total life cycle impact of food products. For most of the 13 environmental indicators reported, plastic packaging is responsible for less than 10% of total life cycle emissions of 23 out of the 30 foods studied (Kan and Miller 2022). Following the European Strategy for plastics, in 2019 the European Commission adopted the Singleuse Plastic Directive. EU rules on single-use plastic products aim to prevent and reduce the impact of certain plastic products on the environment, in particular the marine environment, and on human health. They also aim to promote the transition to a circular economy with innovative and sustainable business models, products and materials. Besides the EPR schemes mentioned before, the Directive comprehends different provisions: an EU-wide ban of specific single use plastics products; a ban of products made from oxo-degradable plastics; consumption reduction of certain single-use plastic items; requirements for beverage bottles; labelling of certain products and awareness raising measures. The Balearic Islands have been one of the pioneers in implementing a circular strategy in plastic packaging management. Through a combination of measures on generic waste generation and specific waste streams such as single-use plastics or food waste, the Ley 8/2019 "De residuos y suelos contaminados de las Illes Balears" offers a strong example of how to adopt a law that has a comprehensive approach to the problem of pollution from waste generation. The main objective of the law is to tackle the problem of waste generation through prevention and improved recycling. It aims to promote waste prevention and reuse through specific measures and binding targets, which also aim to tackle the most problematic and visible waste streams, such as single-use products, plastic packaging and food waste. Even in this innovative law, Extended Producer Responsibility (EPR) schemes put further pressure on producers to fully support the transition.

Another innovative measure has been taken by the French Government in 2020 with the Loi no 105 "Objectifs stratégiques de gestion et de prévention de la production de déchets". The law bans plastic packaging on most fruit and vegetables, indeed from 2021 the country banned plastic straws, cups and cutlery, as well as polystyrene takeaway boxes. A total of 30 types of fruit and vegetables are banned from having any plastic wrapping, including bananas, pears, lemons, oranges, and kiwis. The President of the Republic said the law 105/2020 shows the country's commitment to phase out single use plastics by 2040, however some claims came out because of the short notice and the food waste issue. Philippe Binard, from the European Fresh Produce Association, said that the "removal of plastic packaging from most fruit and vegetables at such short notice does not allow alternatives to be tested and introduced in a timely manner and stocks of existing packaging to be cleared", pointing out the short period of time that is given to the flexible packaging producers to innovate their production.

In April 2022, a Plastic Packaging Tax was introduced by the UK Government. This tax is designed to encourage the use of more recycled plastic and applies to plastic packaging produced in, or imported into, the UK and that does not contain at least 30% recycled plastic. The rate of the Plastic Packaging Tax is set at £200 per tonne of plastic packaging placed onto the UK market. The Tax will apply to businesses that manufacture or import plastic packaging components or import packaged goods into the UK.

Why Dell Reconnect Project Can Also Work for Plastic Packaging

It is clear that the governments are changing the regulatory framework in order to incentivise the circular economy in the plastic industry and avoid single use plastic. However, it is not still clear whether the market is ready for this change and whether it is effectively possible to shift to another material in some applications, as the French case shows with fruit and vegetables. Indeed, the applications in which plastic is involved are various and complex, therefore the sectors and the firms answer differently to the new sustainable demand. A case-by-case analysis should be conducted to evaluate the alternative materials and solutions available in the market for each application.

The transition towards a sustainable production, management and disposal of plastic packaging has already started in Dell Reconnect. However, the evidence shows that the private sector transition is slower. Phelan et al. (2022) conducted a systematic review of 68 corporate sustainability reports to examine how major multinational companies in the food and beverage sector are addressing the environmental issue. According to the research, most companies do not directly acknowledge plastic pollution as an issue. The findings show an insufficient pace of adoption and that the horizon of planning towards sustainable packaging is limited. Only a quarter of the companies are addressing the following key issues: packaging design innovation, increasing effectiveness of recycling, reducing single-use plastics, managing product endof-life processes, and driving stakeholder engagement on plastic waste topics. Approximately half of the companies make broad statements affirming their support for the circular economy, however, many of the same companies maintain that consumer education and increased

Figure 6. Life cycle assessment of plastic

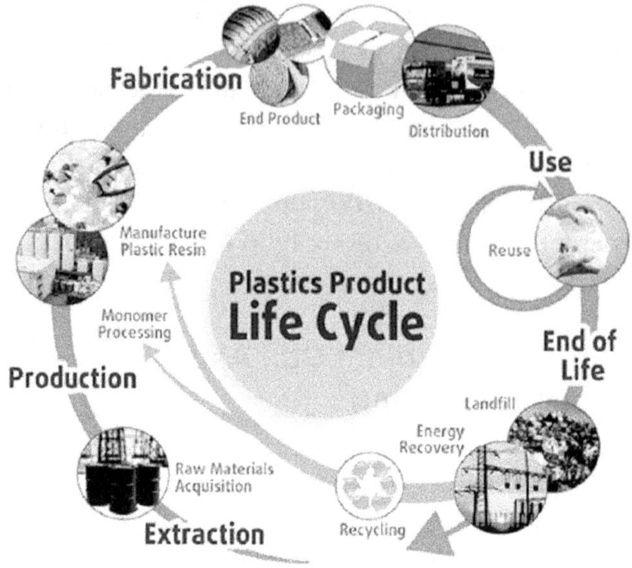

recycling are still their primary focus areas when it comes to plastic waste (Phelan et al., 2022).

During recent years, many studies have been published to review possible non-fossil-based alternatives to BOPP and circular economy strategies to adopt. However, it is difficult to draft a comparative analysis of the alternative non-fossil-based materials (Walker and Rothman 2020), due to methodological problems. Walker and Rothman (2020) review the state-of-the-art in comparative Life Cycle Assessment of fossil-based and bio-based polymers (Figure 6 reports a quick life cycle assessment of plastic). According to the authors, no published articles were found to fully meet the European Union Product Environmental Footprint standards, but the critical review method was used to classify the articles by their level of compliance. 25 articles partially met the PEF standards, giving 39 fossil-based and 50 bio-based polymer case results. In addition, the comparison between the different LCAs was difficult because of the different feedstock sources and processing methods assumed (Walker and Rothman 2020). It was possible to compare seven biobased polymers and seven fossil-based polymers across seven impact categories (energy use, ecotoxicity, acidification, eutrophication, climate change, particulate matter formation and ozone depletion), nevertheless it is not possible to highlight 'best performing' polymers, or to suggest whether fossil-based or bio-based polymers perform best in any impact category due to the methodological problems (Walker and Rothman 2020). The methodological problem brought the European Commission to entrust the Joint Research Centre with a comparative analysis, through the means of life cycle assessment, that assesses the potential environmental impacts of the use of alternative feedstock for plastic films in comparison to using current feedstock. This report is really precious for the literature about flexible packaging, even if it comes from a political organisation, because it is the only one that uses the Product Environmental Footprint (PEF) methodology for all the materials. Within the report, PP film is mostly considered as a reference, being the benchmark that needs to be outperformed. The others non fossil-based polymers considered are Bio-LDPE, which is derived from Sugarcane in Brazil; PLA, which is derived from maize in the U.S.A.; starch-based polymer from corn. The impacts considered, according to the Product Environmental Footprint methodology used are: climate change, ozone depletion, human toxicity (cancer and non-cancer), particulate matter, ionising radiation, photochemical ozone formation, acidification, eutrophication terrestrial, eutrophication freshwater, eutrophication marine, ecotoxicity freshwater, land use, water use, resource use (minerals, metals and fossils). The overall results show that the bio-based LDPE film has averagely the highest impact compared to the other materials, mainly due to the relevant polymer production impacts. Therefore, the use of bio-based ethylene derived from dedicated crop (sugarcane) does not appear to be an environmentally sound alternative.

CO_2-based PP also shows high impacts in many categories. This appears related to high energy consumption in polymer production (e.g., in Climate Change, Ionising Radiation, Resource Use fossils), or chemicals used in CO_2 capture and methanol synthesis. The overall performance of the use of this alternative feedstock for PP production is thus worse compared to its fossil-based counterpart in the majority of the categories. Similar reasoning can be made for the PLA and starch-based polymers. PLA has a higher impact than PP in most of the cases, specifically for what concerns climate change, ozone depletion, particulate matter and land use. At the same time, PLA has less impact on fossil resource use and human toxicity. starch based polymers show higher impacts than PP in climate change, producing more than double the CO_2 kilograms in its lifecycle. Starch-based polymers also have more impacts than PP in ozone depletion, particulate matter and land use, however they show less impact in human toxicity, eutrophication and fossil resource use. Within the non-fossil alternatives available in the market, bioplastics seem to be the most used, but also controversial. This is due to mainly two reasons: methodological problems with the LCAs analysis and an ethical issue that poses in conflict the bio-based plastic production with the agricultural production and sustainability.

A review conducted in 2019 by the Irplast S.p.A. Research and Development in collaboration with the Istituto di Management of the Scuola Superiore Sant'Anna within the Eco.Re.Label project co-financed by Regione Toscana, compares the scientific literature about bioplastics. The characteristics of bioplastics make this product competitive to fossil-based plastics from a theoretical point of view, but the literature highlights controversial facts of bioplastics. In particular, what emerges from the review of four LCAs (Alvarez- Chavez et al (2012), Hottle et al. (2013), Yates et Barlow (2013); Grabowski et al. (2015)) is that bioplastics have several critical points. Firstly, the land occupation for agricultural production, which poses an ethical question. Indeed, occupying the space for the bioplastic production means to not use that space for food production. Secondly, the implementation of industrial agriculture for production with consumption of water, energy and pesticides. Thirdly, the production processes from feedstock to plastics with consumption of energy, water and chemical and biological agents (GMOs). Fourthly, the compostability of biopolymers could be reduced by their mixing with polymers of fossil origin or by the reduced availability of suitable infrastructure for composting. Therefore, from the comparative analysis of the LCAs comes out that the introduction of bioplastics makes it possible to reduce environmental impacts, in particular for GHGs emissions and consumption of non-renewable resources, compared to polymers of fossil origin, however the advantage is less clear for other impact categories such as acidification and eutrophication, which are significantly affected by the processes included in the production phase, both in agriculture and in processing. In addition, the studies which include all phases of the life cycle from cradle to grave, thus including the end-of-life phase, have shown that the choice of the end-of-life scenario of the biopolymer is decisive in the assessment of the overall environmental impacts. Indeed, different end-of-life options, which should be modelled with data of adequate quality, result in different environmental impacts (Bishop et al., 2021). The results shown are confirmed by Kakadellis and Harris (2020) and Mendes and Pedersen (2021). Kakadellis and Harris (2020) clearly state that the LCAs included in their study do not provide enough evidence to state which polymer is best at reducing food waste. Furthermore, the importance of including the food itself in food packaging LCAs is claimed. The environmental impacts of food produce are often significantly higher than those associated with packaging, regardless of the nature of the plastic material (Kakadellis and Harris 2020).

Also, Mendes and Pedersen (2021) state that plastic packaging materials have reduced climate impact, but there are other environmental impacts (e.g. eutrophication, use of water and pesticides, effects on biodiversity) that are less favourable for using bio-based materials and should also be considered (Mendes and Pedersen 2021).

PLASTIC VALUE CHAIN FROM A CIRCULAR ECONOMY PERSPECTIVE

Overview

The plastic circular economy approach that is promoted by the new sustainable European regulatory framework is an alternative to the traditional linear system. The circular economy approach wants to close the loop of the plastic life cycle, increasing also the amount of plastic that is reused or recycled back into the system (Calleja, 2019). A circular plastic economy could contribute to less plastic being downcycled, incinerated and landfilled, and following the direction of making plastic waste a resource for new products in a closed-loop production and consumption system (Johansen et al., 2022). The transition to the circular economy should be made across the entire plastics value chain in order to ensure circular design, production, use and waste management (Johansen et al., 2022).

Despite the circular economy is gaining increasing levels of attention, only 30% of plastic waste is collected for recycling in Europe and most of the waste is downcycled into materials with a lower value than that attributed to the original product (Calleja, 2019). Johansen et al. (2022) made a review of the existing literature about plastic circular economy strategies and gaps, dividing the plastic value chain in six main phases: raw materials, polymer design & production, product production, retailer, use & demand and waste management. According to their review, the downcycling of plastic in the economy is caused by contamination with organic and inorganic matter, and from product designs that combine different materials, like the combination of plastic and paper or aluminium, and thereby inhibit and complicate technical and economically feasible pathways for plastic-waste recycling. Furthermore, the existing knowledge tends to emphasise research focused on the 'end-of-life' phase, aimed at improving recycling and recovery of plastic waste. However, according to Brouwer et al. (2018), the cause of mixed polymers in recycling is not only sorting and recycling processes but the initial design of plastic products and packaging. Packaging and other post-consumer plastics are often designed with multiple polymers, which makes recycling unfeasible from a technical as well as an economic perspective. The results of the literature show an unequal distribution of research across the different phases of the value chain. The majority of published articles focus on the waste-management phase, with a comparatively small number of studies dealing with product design, production, and use. This highlights an important knowledge gap, as most research has been focused on the end-of-life phase, despite increased political and scientific emphasis on the circular economy that, ideally, would include recycling as just one aspect of an overall strategy (Johansen et al., 2022). Indeed, the transition towards a circular economy cannot be achieved solely through changes within the waste-handling system but must be combined with changes in the entire value chain, including the product design and the production phase.

Design Phase

The design is one of the initial phases of the packaging value chain. This phase is extremely important for the implementation of a circular economy strategy as it is outlined by the European Commission with the Extended Producer Responsibility schemes. This stage is crucial because functionality and features of the product are decided, such as colour, recyclability, recipe of the product. Furthermore, the company marketing strategy has a key role in the design phase (Rundh 2016), where the brand has to communicate its product identity, but also the packaging itself. In a circular economy strategy, the packaging itself acquires more importance, even from the consumers point of view. Indeed, the attention of the consumers is now focused also on the environmental impact of the packaging and its recyclability. Especially for food contact applications, the design of flexible packaging is a key phase. As it is pointed out by the regulatory review, food contact plastic applications have to follow precise quality standards and cannot be contaminated by other substances. Indeed, this problem is also reflected in the literature, where a common theme regarding the implementation of circular economy strategy is the challenge of producing packaging with recycled plastics. The main constraints are the high quality standards, but also low migration of toxins, being light-weight and having the ability to keep goods in proper conditions. Many authors assess that recycled plastic can be difficult to use in new products owing to the contamination of polymer types, which may affect aspects such as durability, toxicity and weight.

Production Phase

The production of flexible packaging covers several differentiated stages, from the production of the virgin film towards the reels cutting and the storage. The literature review made by Johansen et al. (2022) shows how only 3 papers out of 60 address this phase of the plastic value chain, highlighting a lack of knowledge in this specific phase. In these papers the main challenge outlined about circular economy strategies is that recycled plastic is most often composed of different polymer types and contaminated with non-plastic materials, such as additives, inks and glues. Therefore, this causes a range of complications, which reduces the value of the final product made by a mix of virgin and recycled plastic (Getor et al., 2020). Indeed, plastic products are designed with mixed polymers and the waste stream becomes, therefore, contaminated, affecting the properties of the recycled plastic (Getor et al., 2020).

This technical problem is a key issue not only for the production phase, but for all the implementation of a plastic circular economy strategy. Indeed, the quality of the feedstock impacts all the plastic value chain and poses the company an important limit, as it has been shown even in the design phase.

HOW TO RESPONSIBLY RECYCLE PLASTIC

Besides the non-fossil-based alternatives, plastic can be recycled through mainly two processes: mechanical recycling and chemical recycling. Mechanical recycling is used widely in Europe and elsewhere, the process involves sorting, washing, shredding and melting of plastic waste to produce granulate and finally new plastic products (Ragaert et al., 2017). Mechanical recycling does not change the chemical structure of the material; however it is a generally environmentally-friendly approach for waste plastic disposal (De Giovanni and Folgiero, 2022). Indeed, substituting virgin plastics with recycled plastics has been proved to be economically and technically viable and environmentally advantageous in many cases. Lazarevic et al. (2010) found that mechanical recycling of waste plastics is more preferable to incineration and landfill, provided that a certain recycled material substitution ratio is achieved.

The cost and the GHG emission of plastic recycling can be further reduced by implementing optimisation strategies, which could make recycled plastics more competitive. Aside from environmental benefits, mechanical plastic recycling is an economically viable practice, as recycled plastics represent a savings of 20–50% in terms of the market prices when compared to virgin counterparts. However, the plastic obtained by mechanical recycling cannot be used for film production in the food and beverage industry because the process can degrade the mechanical properties of plastic materials (Yin et al., 2015). This means that the mixed plastic waste is usually 'down-cycled' into secondary and less valuable products. As the mechanically recycled raw materials are downgraded during the recycling process, this kind of raw materials are not suitable for several manufacturing processes. For example, the sequential or simultaneous stretching extrusion cannot be executed, otherwise the material would be unusable. Further, the efficiency of mechanical recycling is hindered by the difficulty in recovering large amounts of targeted plastic from mixed municipal solid waste. Moreover, it is inefficient for difficult-to-recycle plastics, such as multi-layered or heterogeneous plastic used, for example, for packaging (Singh et al., 2017). In addition, depending on the quality of the incoming waste plastic and the efficiency of the sorting process, residual materials, including plastics, still need to be either incinerated or landfilled. Chemical recycling is the process of converting polymeric waste by changing its chemical structure and turning it back into their functional units that can be used as feedstocks for the production of new virgin like polymers and manufacturing of plastics or other products. There are different chemical recycling technologies, e.g. pyrolysis, gasification, hydro-cracking and depolymerisation. Studies assessing chemical recycling suggested that the climate change impact of pyrolysis and other chemical recycling technologies was significantly lower than that one of incineration. Also, Bora et al. (2020) determined through their study that the chemical recycling pathways such as fast pyrolysis and gasification have the potential to mitigate climate change and other environmental impacts posed by plastic wastes. On the other hand, some LCA studies on pyrolysis based on simulations claims that chemical recycling has a higher environmental impact than mechanical recycling, even if these studies do not consider the difference in the quality of the outputs which could disadvantage options that produce a better quality recycled (Jeswani et al., 2021). Especially for the food and beverage films production, this aspect cannot be excluded from the analysis since it is related to the food quality and safety. Indeed, chemical recycling is nowadays the only process that allows flexible packaging producers to use recycled plastic for food contact films because of its quality and upcycling characteristics. Chemical recycling has also been accepted by the EU Taxonomy as a green practice, however, chemical recycling facilities are not widespread and still have to develop on a large scale.

CONCLUSIONS

This overview of plastic packaging offers insights on its application for Dell Reconnect. The theoretical framework allows one to describe the environmental implications of a circular economy project, the impact in terms of economic sustainability, and the circular strategy to implement. Within Dell Reconnect project, several usages of plastics exist. However, the business framework only focuses on electronics and technology.

Accordingly, business models like Dell Reconnect should be re-engineered to think about plastic as well. Currently, the EU, the single member States and the United Kingdom are innovating their legislations in order to promote circular economy. The new framework involves intensely the flexible packaging world, which is considered with all the plastic industry a key priority for the ecological transition. Indeed, packaging is destined to change rapidly in these years, embracing the sustainable and circular perspective and abandoning just its physical and marketing purpose. Summarising the regulatory innovation that is facing the industry, the three main actions which are impacting the flexible packaging producers are the implementation of extended producer responsibility schemes, the ban of single use plastic and the imposition of a percentage of recycled material content in the products. These three actions indicate the necessity of decreasing virgin fossil-based material, eliminate the unnecessary material and incentivise the recycling of plastic.

Guided by the new legislative requirements, the new market necessities and its policy, Dell Reconnect can implement circular business strategy to reduce the plastic production impacts in all the value chain. The study results indicate that Dell Reconnect can adopt circular economy business model is based on three building blocks: resource recovery, design for recycling, and circular supplies. The resource recovery is obtained through additional uses from resources and to extract more value from them by avoiding final disposal for as long as possible. The regeneration process through the regranulation of the film waste can give new life to the film waste. The design for recycling wants to follow the extended producer responsibility schemes, designing products that are easier to recycle. As the results show, the four key actions

Figure 7. Drivers influencing the circular economy of plastic

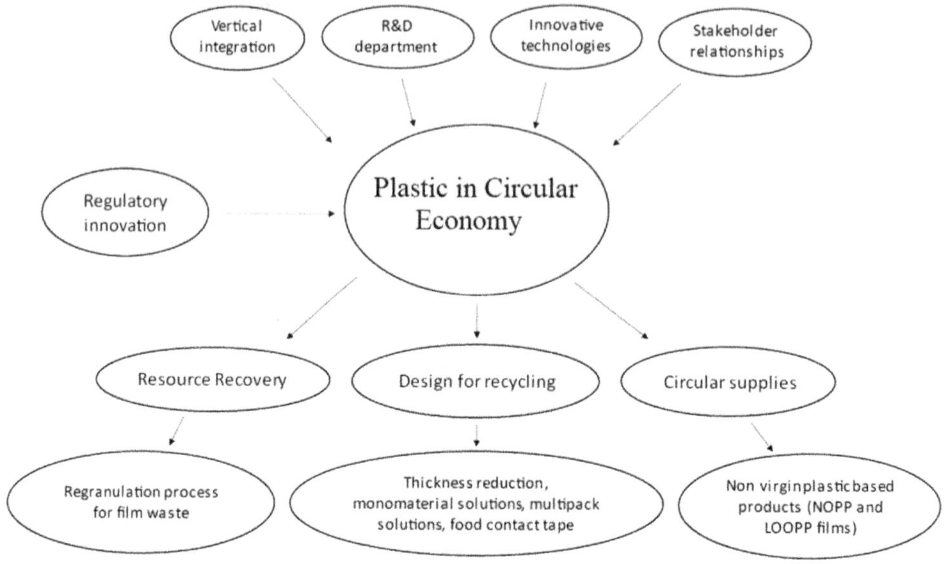

to be undertaken by DellReconnect are the thickness reduction, the mono material solutions, the multipack solutions and material reduction in contact packaging. The Circular Supplies business block goal is to lessen dependence on virgin fossil-based resources. DellReconnect can be a pioneer in the electronics sector also recycling plastic.

Furthermore, the results show that, in order to implement this business model, Dell Reconnect can be clearly organized differently. First of all, the Company vertical integration structure. Dell Reconnect takes advantage of the industrial symbiosis between the Dell and Goodwill sites which fosters eco-innovation, creates and share knowledge, improves business and technical processes (Maranesi and De Giovanni, 2020). The vertical integration brings benefits to the Company in terms of cost reduction, increased revenues and competitive advantage. Secondly, Dell Reconnect can invest in innovative technologies both for the product and process development, in order to increase the efficiency and reduce environmental impacts. In fact, innovative technologies can foster the circular economy through efficiency and transparency (Maranesi and De Giovanni, 2020). Thirdly, the partnership can be developed as a qualified Research and Development department that collaborates with the production in developing tailor-made innovative solutions. The R&D department plays a strategic role in developing the tailor-made products for customers, which continuously adapt their demand according to the market requests. Eventually, another important factor to achieve these results has been the relationship with the stakeholders. Indeed, the relationship between all the actors of the value chain is the key factor in order to implement a circular economy strategy. The relational dimension of circular economy is fundamental to pursue mutually supporting progress and a higher level of competitiveness, especially within the supply chain members. The collaboration and interaction with suppliers and customers lead to superior operational and environmental performances. Dell Reconnect can be a virtuous circular business model that embraces these new regulations and changes its production and process towards a more sustainable perspective. The circular business model adopted can represent a benchmark for the flexible packaging producers in facing the transition and the difficulties in implementing the circular economy. The case study overcomes the design and production gaps found by Johansen et al. (2022) and represents a virtuous case in the plastic private sector that is struggling to implement a concrete circular economy strategy (Phelan et al., 2022). Firstly, our research demonstrates that there might be alternative feedstocks to virgin plastics, which overcome the problems claimed by the literature.

Furthermore, DellReconnect demonstrates that the design and production phases might play a key role in reducing the impacts themselves through new efficient technology, waste recovery and innovative design, but also they can increase the waste recycling efficiency at the end of the value chain through the products development thanks to the R&D department quality. Indeed, the research demonstrates that investments in new technologies, investments in high quality human resources (especially at Goodwill's sites), a high qualified Research and Development department that collaborates with the production in innovating its design and production value chain phases are the solid base in implementing the ecological transition. In addition, the vertical integration plays a key role in accelerating the internal development in the production and the launch to the market. Eventually, the case highlights the need for a strong stakeholder relationship to embrace a holistic perspective as required by Johansen et al. (2022). Indeed, the relationship with all the actors in the value chain are key to communicate the needs and requirements for the implementation of a circular economy. From the feedstock suppliers to the waste managers, the virtuous relationships between the stakeholders are crucial in order to optimise investments in research and technologies, and design recyclable products. In this sense, the prior commercial relationship becomes a really valuable asset in the new circular economy business model, where the dialogue and transparent communication with the other stakeholders is needed more than before.

REFERENCES

Al-Maaded, M., Madi, N. K., Kahraman, R., Hodzic, A., & Ozerkan, N. G. (2012). An overview of solid waste management and plastic recycling in Qatar. *Journal of Polymers and the Environment*, *20*(1), 186–194. doi:10.100710924-011-0332-2

Álvarez-Chávez, C. R., Edwards, S., Moure-Eraso, R., & Geiser, K. (2012). Sustainability of bio-based plastics: General comparative analysis and recommendations for improvement. *Journal of Cleaner Production*, *23*(1), 47–56. doi:10.1016/j.jclepro.2011.10.003

Andreasi Bassi, S., Boldrin, A., & Faraca, G. (2020). Extended producer responsibility: How to unlock the environmental and economic potential of plastic packaging waste? *Resources Conservation and Recycling*.

Beers, K., Schumacher, K., Migler, K., Morris, K. C., & Kneifel, J. (2022). *An Assessment of Mass Balance Accounting Methods for Polymers Workshop Report*. NIST Special Publication, 1500, 206.

Bishop, G., Styles, D., & Lens, P. N. (2021). Environmental performance comparison of bioplastics and petrochemical plastics: A review of life cycle assessment (LCA) methodological decisions. *Resources, Conservation and Recycling*, *168*, 105451. doi:10.1016/j.resconrec.2021.105451

Bora, R. R., Wang, R., & You, F. (2020). Waste polypropylene plastic recycling toward climate change mitigation and circular economy: Energy, environmental, and technoeconomic perspectives. *ACS Sustainable Chemistry & Engineering*, *8*(43), 16350–16363. doi:10.1021/acssuschemeng.0c06311

Brouwer, M. T., van Velzen, E. U. T., Augustinus, A., Soethoudt, H., De Meester, S., & Ragaert, K. (2018). Predictive model for the Dutch post-consumer plastic packaging recycling system and implications for the circular economy. *Waste Management, 71*, 62-85.

Calleja, D. (2019). Why the "New Plastics Economy" must be a circular economy. Field Actions Science Reports. *The Journal of Field Actions*, (19), 22-27.

Coates, G. W., & Getzler, Y. D. (2020). Chemical recycling to monomer for an ideal, circular polymer economy. *Nature Reviews. Materials*, *5*(7), 501–516. doi:10.103841578-020-0190-4

De Giovanni, P. (2016). Coordination in a distribution channel with decisions on the nature of incentives and share-dependency on pricing. *The Journal of the Operational Research Society*, *67*(8), 1034–1049. doi:10.1057/jors.2015.118

De Giovanni, P. (2019). When feature-based production capabilities challenge operations: Drivers, moderators, and performance. *International Journal of Operations & Production Management*, *40*(2), 221–242. doi:10.1108/IJOPM-04-2019-0309

De Giovanni, P. (2020). An optimal control model with defective products and goodwill damages. *Annals of Operations Research*, *289*(2), 419–430. doi:10.100710479-019-03176-4

De Giovanni, P. (2021). *Dynamic quality models and games in digital supply chains*. Springer Nature. doi:10.1007/978-3-030-66537-1

De Giovanni, P. (2022). Leveraging the circular economy with a closed-loop supply chain and a reverse omnichannel using blockchain technology and incentives. *International Journal of Operations & Production Management*, *42*(7), 959–994. doi:10.1108/IJOPM-07-2021-0445

De Giovanni, P., & Folgiero, P. (2022). Circular Economy Strategies and the Emerging Trend of Circular Districts. *Routledge-Giapichelli*.

De Giovanni, P., & Ramani, V. (2017). Product cannibalization and the effect of a service strategy. *The Journal of the Operational Research Society*, 1–17.

De Giovanni, P., & Vinzi, V. E. (2014). The benefits of a monitoring strategy for firms subject to the Emissions Trading System. *Transportation Research Part D, Transport and Environment*, *33*, 220–233. doi:10.1016/j.trd.2014.06.008

De Giovanni, P., & Zaccour, G. (2022). A selective survey of game-theoretic models of closed-loop supply chains. *Annals of Operations Research*, *314*(1), 1–40. doi:10.100710479-021-04483-5

Delft, C. E. (2020). *Exploration chemical recycling–Extended summary. What is the potential contribution of chemical recycling to Dutch climate policy*. Academic Press.

Duhaylongsod, J. B., & De Giovanni, P. (2018). The impact of innovation strategies on the relationship between supplier integration and operational performance. *International Journal of Physical Distribution & Logistics Management, 49*(2), 156–177. doi:10.1108/IJPDLM-09-2017-0269

European Environmental Agency. (2021). *Plastics, the circular economy and Europe¢s environment — A priority for action.* Publications Office of the European Union.

Fernqvist, F., Olsson, A., & Spendrup, S. (2015). What's in it for me? Food packaging and consumer responses, a focus group study. *British Food Journal, 117*(3), 1122–1135. doi:10.1108/BFJ-08-2013-0224

Gear, M., Sadhukhan, J., Thorpe, R., Clift, R., Seville, J., & Keast, M. (2018). A life cycle assessment data analysis toolkit for the design of novel processes–A case study for a thermal cracking process for mixed plastic waste. *Journal of Cleaner Production, 180*, 735–747. doi:10.1016/j.jclepro.2018.01.015

Getor, R. Y., Mishra, N., & Ramudhin, A. (2020). The role of technological innovation in plastic production within a circular economy framework. *Resources, Conservation and Recycling, 163*, 105094. doi:10.1016/j.resconrec.2020.105094

Geueke, B., Groh, K., & Muncke, J. (2018). Food packaging in the circular economy: Overview of chemical safety aspects for commonly used materials. *Journal of Cleaner Production, 193*, 491–505. doi:10.1016/j.jclepro.2018.05.005

Geuke, B., Groh, K., & Muncke, J. (2018). Food Packaging in the circular economy: Overview of chemical safety aspects for commonly used materials. Elsevier. *Journal of Cleaner Production, 193*, 491–505. doi:10.1016/j.jclepro.2018.05.005

Grabowski, A., Selke, S. E., Auras, R., Patel, M. K., & Narayan, R. (2015). Life cycle inventory data quality issues for bioplastics feedstocks. *The International Journal of Life Cycle Assessment, 20*(5), 584–596. doi:10.100711367-015-0853-3

Gu, F., Guo, J., Zhang, W., Summers, P. A., & Hall, P. (2017). From waste plastics to industrial raw materials: A life cycle assessment of mechanical plastic recycling practice based on a real-world case study. *The Science of the Total Environment, 601*, 1192–1207. doi:10.1016/j.scitotenv.2017.05.278 PMID:28605837

Gu, F., Hall, P., & Miles, N. J. (2016). Development of composites based on recycled polypropylene for injection moulding automobile parts using hierarchical clustering analysis and principal component estimate. *Journal of Cleaner Production*, *137*, 632–643. doi:10.1016/j.jclepro.2016.07.028

Gu, F., Hall, P., & Miles, N. J. (2016). Performance evaluation for composites based on recycled polypropylene using principal component analysis and cluster analysis. *Journal of Cleaner Production*, *115*, 343–353. doi:10.1016/j.jclepro.2015.12.062

Gu, F., Hall, P., Miles, N. J., Ding, Q., & Wu, T. (2014). Improvement of mechanical properties of recycled plastic blends via optimizing processing parameters using the Taguchi method and principal component analysis. *Materials & Design*, *62*, 189-198.

Hottle, T. A., Bilec, M. M., & Landis, A. E. (2017). Biopolymer production and end of life comparisons using life cycle assessment. *Resources, Conservation and Recycling*, *122*, 295–306. doi:10.1016/j.resconrec.2017.03.002

Jalali, H., Ansaripoor, A., Ramani, V., & De Giovanni, P. (2022). Closed-loop supply chain models with coopetition options. *International Journal of Production Research*, *60*(10), 3078–3106. doi:10.1080/00207543.2021.1910871

Jalali, H., Ansaripoor, A. H., & De Giovanni, P. (2020). Closed-loop supply chains with complementary products. *International Journal of Production Economics*, *229*, 107757. doi:10.1016/j.ijpe.2020.107757

Jeswani, H., Krüger, C., Russ, M., Horlacher, M., Antony, F., Hann, S., & Azapagic, A. (2021). Life cycle environmental impacts of chemical recycling via pyrolysis of mixed plastic waste in comparison with mechanical recycling and energy recovery. *The Science of the Total Environment*, *769*, 144483. doi:10.1016/j.scitotenv.2020.144483 PMID:33486181

Johansen, M. R., Christensen, T. B., Ramos, T. M., & Syberg, K. (2022). A review of the plastic value chain from a circular economy perspective. *Journal of Environmental Management*, *302*, 113975. doi:10.1016/j.jenvman.2021.113975 PMID:34700081

Kakadellis, S., & Harris, Z. M. (2020). Don't scrap the waste: The need for broader system boundaries in bioplastic food packaging life-cycle assessment–A critical review. *Journal of Cleaner Production*, *274*, 122831. doi:10.1016/j.jclepro.2020.122831

Kan, M., & Miller, S. A. (2022). Environmental impacts of plastic packaging of food products. *Resources, Conservation and Recycling*, *180*, 106156. doi:10.1016/j.resconrec.2022.106156

Kozderka, M., Rose, B., Kočí, V., Caillaud, E., & Bahlouli, N. (2016). High impact polypropylene (HIPP) recycling–Mechanical resistance and Lifecycle Assessment (LCA) case study with improved efficiency by preliminary sensitivity analysis. *Journal of Cleaner Production*, *137*, 1004–1017. doi:10.1016/j.jclepro.2016.07.217

Lazarevic, D., Aoustin, E., Buclet, N., & Brandt, N. (2010). Plastic waste management in the context of a European recycling society: Comparing results and uncertainties in a life cycle perspective. *Resources, Conservation and Recycling*, *55*(2), 246–259. doi:10.1016/j.resconrec.2010.09.014

Maddah, H. A. (2016). Polypropylene as a promising plastic: A review. *American Journal of Political Science*, *6*(1), 1–11.

Maranesi, C., & De Giovanni, P. (2020). Modern circular economy: Corporate strategy, supply chain, and industrial symbiosis. *Sustainability*, *12*(22), 9383. doi:10.3390u12229383

Matthews, C., Moran, F., & Jaiswal, A. K. (2021). A review on European Union's strategy for plastics in a circular economy and its impact on food safety. *Journal of Cleaner Production*, *283*, 125263. doi:10.1016/j.jclepro.2020.125263

Mendes, A. C., & Pedersen, G. A. (2021). Perspectives on sustainable food packaging:–is bio-based plastics a solution? *Trends in Food Science & Technology*, *112*, 839–846. doi:10.1016/j.tifs.2021.03.049

Nessi, S., Bulgheroni, C., Konti, A., Sinkko, T., Tonini, D., & Pant, R. (2018). *Environmental sustainability assessment comparing through the means of lifecycle assessment the potential environmental impacts of the use of alternative feedstock (biomass, recycled plastics, CO2) for plastic articles in comparison to using current feedstock (oil and gas).* Draft Report for Stakeholder Consultation (Part 2).

Nielsen, T. D., Hasselbalch, J., Holmberg, K., & Stripple, J. (2020). Politics and the plastic crisis: A review throughout the plastic life cycle. *Wiley Interdisciplinary Reviews. Energy and Environment*, *9*(1), e360. doi:10.1002/wene.360

Otto, S., Strenger, M., Maier-Nöth, A., & Schmid, M. (2021). Food packaging and sustainability–Consumer perception vs. correlated scientific facts: A review. *Journal of Cleaner Production, 298*, 126733. doi:10.1016/j.jclepro.2021.126733

Phelan, A. A., Meissner, K., Humphrey, J., & Ross, H. (2022). Plastic pollution and packaging: Corporate commitments and actions from the food and beverage sector. *Journal of Cleaner Production, 331*, 129827. doi:10.1016/j.jclepro.2021.129827

Ragaert, K., Delva, L., & Van Geem, K. (2017). Mechanical and chemical recycling of solid plastic waste. *Waste Management, 69*, 24-58.

Rahim, R., & Raman, A. A. A. (2017). Carbon dioxide emission reduction through cleaner production strategies in a recycled plastic resins producing plant. *Journal of Cleaner Production, 141*, 1067–1073. doi:10.1016/j.jclepro.2016.09.023

Rajendran, S., Hodzic, A., Scelsi, L., Hayes, S., Soutis, C., Alma'Adeed, M., & Kahraman, R. (2013). Plastics recycling: Insights into life cycle impact assessment methods. *Plastics, Rubber and Composites, 42*(1), 1–10. doi:10.1179/1743289812Y.0000000002

Ramani, V., & De Giovanni, P. (2017). A two-period model of product cannibalization in an atypical Closed-loop Supply Chain with endogenous returns: The case of DellReconnect. *European Journal of Operational Research, 262*(3), 1009–1027. doi:10.1016/j.ejor.2017.03.080

Rundh, B. (2016). The role of packaging within marketing and value creation. *British Food Journal, 118*(10), 2491–2511. doi:10.1108/BFJ-10-2015-0390

Sacco, A., & De Giovanni, P. (2019). Channel coordination with a manufacturer controlling the price and the effect of competition. *Journal of Business Research, 96*, 97–114. doi:10.1016/j.jbusres.2018.09.001

Sheriff, K. M., Subramanian, N., Rahman, S., & Jayaram, J. (2017). Integrated optimization model and methodology for plastics recycling: Indian empirical evidence. *Journal of Cleaner Production, 153*, 707–717. doi:10.1016/j.jclepro.2016.07.137

Singh, N., Hui, D., Singh, R., Ahuja, I. P. S., Feo, L., & Fraternali, F. (2017). Recycling of plastic solid waste: A state of art review and future applications. *Composites. Part B, Engineering, 115*, 409–422. doi:10.1016/j.compositesb.2016.09.013

Sommerhuber, P. F., Wang, T., & Krause, A. (2016). Wood–plastic composites as potential applications of recycled plastics of electronic waste and recycled particleboard. *Journal of Cleaner Production, 121*, 176–185. doi:10.1016/j.jclepro.2016.02.036

Somoza-Tornos, A., Gonzalez-Garay, A., Pozo, C., Graells, M., Espuña, A., & Guillén-Gosálbez, G. (2020). Realizing the potential high benefits of circular economy in the chemical industry: Ethylene monomer recovery via polyethylene pyrolysis. *ACS Sustainable Chemistry & Engineering, 8*(9), 3561–3572. doi:10.1021/acssuschemeng.9b04835

Steenis, N. D., Van Herpen, E., Van Der Lans, I. A., Ligthart, T. N., & Van Trijp, H. C. (2017). Consumer response to packaging design: The role of packaging materials and graphics in sustainability perceptions and product evaluations. *Journal of Cleaner Production, 162*, 286–298. doi:10.1016/j.jclepro.2017.06.036

Stichnothe, H., & Azapagic, A. (2013). Life cycle assessment of recycling PVC window frames. *Resources, Conservation and Recycling, 71*, 40–47. doi:10.1016/j.resconrec.2012.12.005

The European House, Ambrosetti. (2013). *L'eccellenza della filiera della plastica per il rilancio industriale dell'Italia e dell'Europa*. Author.

Trabucco, M., & De Giovanni, P. (2021). Achieving Resilience and Business Sustainability during COVID-19: The Role of Lean Supply Chain Practices and Digitalization. *Sustainability, 13*(22), 12369. doi:10.3390u132212369

UNEP. (2020). *Plastic Waste – Background Report*. UNEP.

Vishkaei, B. M., Fathi, M., Khakifirooz, M., & De Giovanni, P. (2021). Bi-objective optimization for customers' satisfaction improvement in a Public Bicycle Sharing System. *Computers & Industrial Engineering, 161*, 107587. doi:10.1016/j.cie.2021.107587

Wäger, P. A., & Hischier, R. (2015). Life cycle assessment of post-consumer plastics production from waste electrical and electronic equipment (WEEE) treatment residues in a Central European plastics recycling plant. *The Science of the Total Environment*, *529*, 158–167. doi:10.1016/j.scitotenv.2015.05.043 PMID:26022405

Walker, S., & Rothman, R. (2020). Life cycle assessment of bio-based and fossil-based plastic: A review. *Journal of Cleaner Production*, *261*, 121158. doi:10.1016/j.jclepro.2020.121158

White, A., & Lockyer, S. (2020). Removing plastic packaging from fresh produce–what's the impact? *Nutrition Bulletin*, *45*(1), 35–50. doi:10.1111/nbu.12420

Yates, M. R., & Barlow, C. Y. (2013). Life cycle assessments of biodegradable, commercial biopolymers—A critical review. *Resources, Conservation and Recycling*, *78*, 54–66. doi:10.1016/j.resconrec.2013.06.010

Yin, S., Tuladhar, R., Shi, F., Shanks, R. A., Combe, M., & Collister, T. (2015). Mechanical reprocessing of polyolefin waste: A review. *Polymer Engineering and Science*, *55*(12), 2899–2909. doi:10.1002/pen.24182

Chapter 11
Blockchain for Circular Economy in the Furniture Sector:
The Case of Cubo Design S.r.l.

Valentina Arangiaro
Simest S.p.A., Italy

Behzad Maleki Vishkaei
Luiss University, Italy

Pietro De Giovanni
https://orcid.org/0000-0002-1247-4807
Luiss University, Italy & Luiss-X.ITE Research Center, Italy

EXECUTIVE SUMMARY

The research aims at examining the potential use of blockchain technology in the furniture sector's supply chain. To better understand how the furniture sector is articulated, Cubo design S.r.l and its new 4.0 production plant are used as a case study. Cubo design is an SME from Abruzzo active in the manufacture of parts and accessories of furniture. The authors describe the switch the furniture sector has undergone from "old" handicraft production with a high presence of labor and with a significant presence of external suppliers and manual production methods to the new paradigm based on smart factories. From this point, they evaluate the possible and future adoption of blockchain technologies in terms of the benefits the technology could bring to the sector, how it can be combined with 4.0 logic, and what improvements it would create to support the furniture circular economy.

DOI: 10.4018/978-1-6684-5001-7.ch011

INTRODUCTION

In the last decades, consumers' awareness regarding ethical and environmental issues has increased, thanks to a growingly interconnected and constantly updated world. The European Circular Economy Action Plan identifies circular economy (CE) as an essential element of a broader transformation towards climate neutrality and long-term competitiveness. The main environmental benefits obtained by the companies are resource recovery from waste regeneration and exploitation, resource efficiency, circular inputs, and the reduction in CO2 emissions (Maranesi & De Giovanni, 2020).

In the furniture sector, with the developing technology, manufacturers turn to customer-oriented production models. So, a traceable furniture chain prototype that develops production focused on customer demands is proposed as an exemplary scenario (Baygin et al., 2020). It is believed that the companies in the wooden furniture manufacturing sector need to react and take action for improving their environmentally related practices and enhance their image as ''green companies'' in terms of energy saving and wood waste reduction (Daian & Ozarska, 2009). In general, economies of scale and economic incentives are needed to make repair and refurbishment viable. Currently there are weak drivers and underinvestment in the collection and logistics for furniture take-back. Producer responsibility mechanisms are not widely used in the furniture sector (Furn, 2022). As wood is one of the main raw materials being used in this sector and one of the current challenges in the world is deforestation, implementing new technologies that can help this industry to manage wood consumption via reusing, recycling, and controlling the source of the initial materials is a vital step in terms of circular economy.

Based on a report that is published by BioReg in 2020, the wood waste value chain includes steps from generation of wood waste to its valorisation, including transport, collection, treatment and processing. Figure 1 shows the numerous actors are concerned throughout the whole value chain.

Blockchain technology ensures better acceptance among consumers when purchasing refurbished goods, and among firms when using recycled materials (De Giovanni, 2022). Blockchain technology provides a platform for recordkeeping and data-sharing. Digital ledgers, containing records and transactions, are visible to all authorised participants of a network. This visibility increases transparency and traceability (Francisco & Swanson, 2018). Some common benefits of implementing blockchain technology include

Figure 1. Actors that are concerned throughout the whole wood waste value chain (www.BioReg.eu)

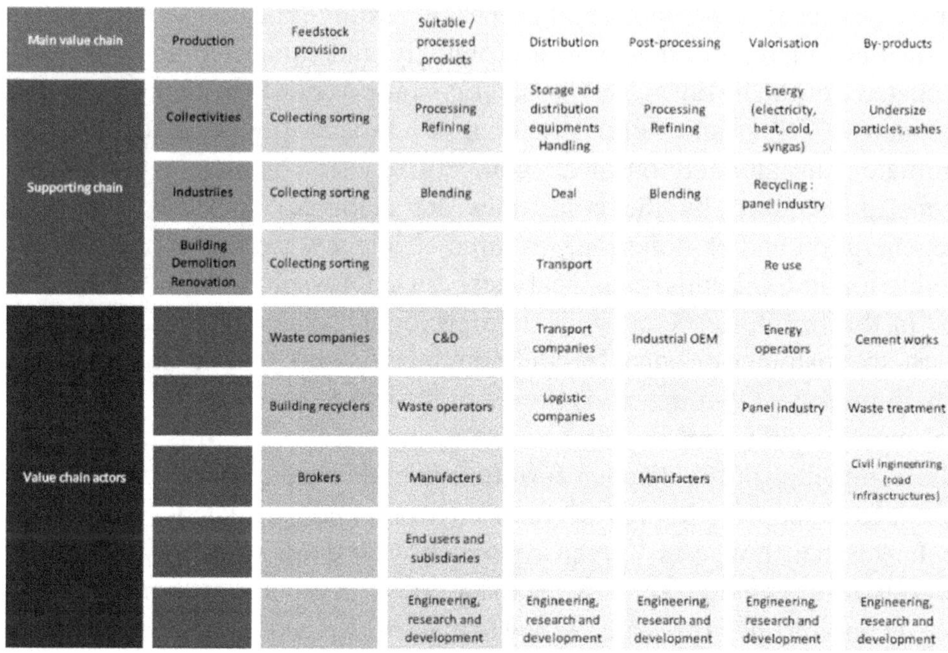

delivering a transparent, decentralized, secure transaction process that may regenerate resources, reduce costs, and improve efficiency and responsiveness. These characteristics, in the long run, yield sustainable economic, social and environmental influences in the circular economy (Kouhizadeh et al., 2020). This chapter aims to find out how blockchain technology can be helpful in the furniture sector to enhance traceability, information transparency, and circular economy practices.

Blockchain and Traceability

The traceability issues limit the consumers' perception of product quality and make it impossible for consumers to follow the purchased goods during the entire delivery path. The management of SCs in global contexts challenge firms not only in terms of traceability but also in terms of high transaction costs (Naclerio and De Giovanni, 2022). These two drawbacks call for the implementation of blockchain technology (Biswas et al., 2022). Traceability in the supply chain struggles with data fragmentation and centralized controls, making it more vulnerable to both data modification and management (Prencipe et al., 2022). The traditional traceability system using the Internet of Things (IoT) is being frequently employed because it can control and store specific information in all four stages (production, processing, distribution, and consumption) in the supply chain. But it is still based on the centralized server-client paradigm (Feng et al., 2020) meaning that stakeholders and consumers have to rely on a single information point.

Blockchain technology is becoming increasingly relevant as a traceability tool, especially in the field of supply chain management. The concept of Blockchain was firstly introduced in the year 2008 by a person under the pseudonym Satoshi Nakamoto in correlation with the topic of the cryptocurrency Bitcoin. As affirmed by Risius et al. (2017) in its generic form, blockchain technology refers to a fully distributed system for cryptographically capturing and storing a consistent, immutable, linear event log of transactions between networked actors. The term Blockchain itself explains the way it stores transactional data which is a sequence of blocks that are linked in the form of a chain (Agrawal et al., 2021). In the field of supply chain management, this technology can enhance transparency, authenticity, trust, and security and could create a real informed environment where all intermediaries are removed.

Few solutions have been proposed to solve the traceability issue and improve transparency and security, but the most promising one seems blockchain technology, thanks to its enabler but complementary nature in terms of innovation (De Giovanni, 2020a). A study conducted by Galvez et al. (2018) on the application of the technology highlights the benefits that can be derived from it and proposes an example of a plant food production chain, where smart contracts track all the production information from seeding procurement to consumption. The information exchange takes place from beginning to end in a very transparent way, and once it is transmitted it cannot be altered. For stakeholders of the industry, creating a traceable and transparent supply chain, with real-time information "builds better relationships, increases efficiency, and reduces the risk and cost of product recalls, counterfeit, and unethical labor (Hader et al., 2022). The adoption of a blockchain-based traceability system is one of the possible solutions to traceability challenges and ensures seamless information sharing among supply-chain partners (Agrawal et al., 2021). It is necessary for all supply chain participants to be completely informed about the product life cycle (PLC) to guarantee products' safety, sustainability, and value (De Giovanni, 2019a). Moreover, the Supply chain can be under the spotlight for intolerable and unethical production practices, child labor, imbalanced wages, incorrect discharge of toxic waste, and illicit trades (Agrawal et al. 2018). It is essential for customers to access product data that can facilitate ethical buying practices or assure product authenticity (Agrawal et al., 2021).

The luxury sector has always been impacted by the weight of counterfeits and the part of the products in the furniture sector are considered luxury goods. Since fake products are becoming more and more sophisticated and hard to distinguish from real ones, "one of the major challenges for luxury goods manufacturers is to assure customers that they are buying a genuine product" (Schmidt et al., 2021). Consequently, blockchain technology represents the perfect trade-off to re-introduce trust and transparency and advance the quality and the speed of the supply chain (Kshetri, 2018). In this regard, "the blockchain traceability anti-counterfeiting platform" (Zhuo et al., 2020) is a cutting-edge project, thanks to the collaboration between manufacturing and transportation enterprises. The platform uses state sensors and RFID technologies to aggregate product information, which is then stored in the blockchain system (De Boissieu et al., 2021). However, it is fundamental for luxury brands to follow the government's directives concerning the employment of blockchains and counterfeiting. Even if several solutions have been introduced to solve these issues, "further efforts are needed to develop highly scalable blockchains that respond to the needs of all food supply chain stakeholders" (Rejeb et al., 2020).

In recent years, the furniture industry has undergone major changes in terms of innovation and digitalization, which have enabled the sector to grow exponentially and become one of the cornerstones of the world economy. In 2021, despite the pandemic crisis, the furniture sector has reached a total revenue of US$673.0 billion worldwide, doubling its performance from 2000. The sector is distributed into seven segments: Living-room Furniture, Bedroom Furniture, Kitchen & Dining Furniture, Outdoor Furniture, Home Office Furniture, Lamps & Lighting, and Floor Covering. The largest in terms of revenue is the Livingroom furniture segment, which accounts for US$232.5 billion (35% of the total), followed by the Bedroom and Kitchen segments (Statista.com).

In the furniture sector, the supply chain is very complex and composed of several different semifinished products (which need to be assembled), consequently, manufacturers have to interface with a great number of suppliers. The high fragmentation of the sector into thousands of small production units and the presence of strong district-related economies, located in every part of the globe, make it even more challenging to create solid and transparent

Figure 2. Example of Supply Chain for furniture products (Gonczol et al., 2020)

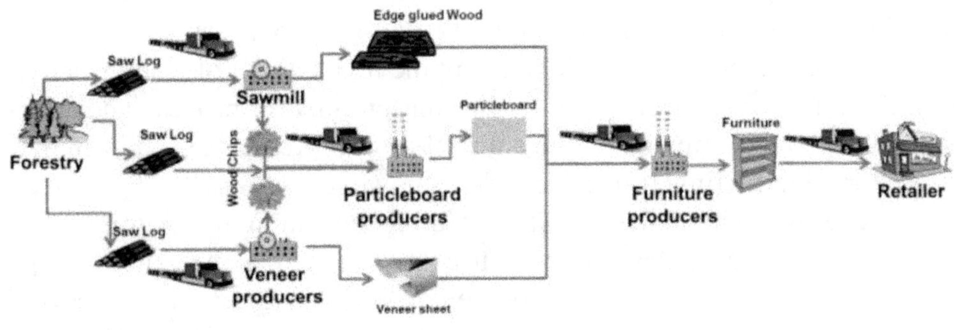

business relationships (De Giovanni, 2019a). Figure 2 shows an example of the supply chain in the proposed sector. Moreover, the spread of Covid-19 and the recent developments linked to the Russo-Ukrainian conflict have contributed to aggravating trade relations and causing supply chain disruption and breakage. As never before, the demand for secure traceability, and reliability systems on the product and its suppliers exponentially increased, especially for what concerns the wood, and the metal supply chains.

Solving these issues requires a distributed database in which the participants of each supply chain can store data in chronological order, through a distributed consensus protocol, and where real-time information is provided, improving the feasibility and reliability of information tracing. The info tracing tracking system should combine the information related to the product quality with those linked to the traceability, through physical and digital documents, in a safe online information system. To date, the available technology that meets these requirements is blockchain, which has been widely considered the key practical solution to supply chain management issues.

As we mentioned before blockchain is a decentralized, distributed and immutable ledger comprised of a cryptographically linked chain of record collection (Hewa et al., 2021) and one of its appealing characteristics is smart contracts. The latter will help all applicants to gain considerable benefits and to form an exclusive transparent and trusted economic environment, as well as ensure the complete security of the whole supply chain. blockchain could be also intended as an enabler that leads existing manufacturing information systems, such as enterprise resource planning (ERP) and manufacturing execution system (MES), which are particularly essential in this type of sector.

In particular, Considering the circular economy aspect, blockchain technology helps firms record all transactions in a chain, in which each node contains several types of information related to the transactions. Firms can use blockchain technologies to document the real amount of resources used as well as the wasted materials during the production process and, more generally, give information on the entire flow of ingredients used from their origin through the final usage phase (De Giovanni, 2020b).

To better understand the potentiality of blockchain in this sector, we will study the wood supply chain, since its use has gained great relevance in terms of contribution to climate and environmental issues. In Europe, two important certifications promote sustainable management of the planet's forest heritage: Forest Stewardship Council (FSC) and Program for the Endorsement of Forest Certification (PEFC).

According to Appelhanz (2013), the wood supply chain is composed of 4 steps: forestry, semi-finished products, finished products, and final products. The first step is extremely significant not only because it is the starting point of the supply chain and the next steps strictly depend on it, but above all, also because in this phase the wood is harvested, and it is very important to avoid any kind of wood waste and maintain the best possible economic use of the forested land. Considering the circular economy concept, the application of a traceable system as blockchain would increase resource efficiency and capacity utilization, decrease damage to the ecosystem, as well as avoid theft of wood and illegal felling.

The logistics of semi-finished goods manufacturers are also crucial, and it needs to be improved through blockchain technology in order to prevent redundant flows and input data errors. blockchain will also be essential to capture the quality of the wood and monitor unexpected events (e.g., discoloration of the timber body). This process can be carried out with the help of an RFID transponder to be placed inside the log, which will record and store the history of the log and will periodically notify the ERP or EM systems.

Another advantage of the application of blockchain concerns law and regulations. Since governments' directives are frequently changing and can also vary from country to country, the technology can inform all the participants of the supply chain about the presence of new standards and regulations to be met (e.g., Formaldehyde emissions are strictly regulated both via community regulations and national laws enforced by several member states of the EU). Table (1) summarizes the benefits given using blockchain during the manufacturing stages of processed and semi-finished products.

Table 1. Improvements using blockchain technology

Business process	Optimization through Blockchain
Raw materials receipt	Simplification of assortment based on captured information about quality and quantity, automated creation of documents
Storage	Control of storage process, Efficient storage management
Manufacturing	Monitor the process, Ensure compliance with standards in the production
Logistics	Prevent redundant flows and input data errors

Accordingly, our first research question is devised as follows to investigate how blockchain technology can help the furniture sector by enhancing information transparency.

RQ 1. How can blockchain technology guarantee information transparency in the furniture sector?

Blockchain and Industry 4.0

For some years now, the furniture sector has been facing a digital revolution in which, the expansion of communications, wireless networks, and the development of increasingly intelligent robots and technological machinery have changed products, services, and production methods. In the new competitive environment, producing pieces, often one-of-a-kind pieces, with traditional tools and obsolete industrial machinery, has become not sustainable anymore; therefore, a paradigm shift has been made to improve efficiency and productivity.

"Industry 4.0" has proven to be the best solution to the structural problems that have gripped the wood-furniture sector. Launched by the German Government in 2011, Industry 4.0 usually refers to the fourth industrial revolution (Cagle et al., 2020). It defines the organization of production processes based on technology and devices that communicate with each other: a model of the smart factory of the future, where computer-driven systems monitor physical processes and make decentralized decisions based

on self-organizing mechanisms. Therefore, "Industry 4.0" is a technological paradigm of rupture with the past and a tool for management to renew and innovate their companies with an eye toward the future (Kagermann et al., 2011).

Cyber-Physical Systems, Internet of things (IoT), big data analytics, cloud manufacturing, and fog computing are the key enabling technologies of Industry 4.0. They allowed the creation of a flexible and customizable production with short production times, the achievement of an even more organized production - based on realistic forecasts and without the need for stock in the warehouse ("just in time" production) -, and the integration not only of commercial partners but also of users in one network (Duhaylongsod and De Giovanni, 2018).

One interesting example of 4.0 services related to the furniture sector is "Sophia", the IoT platform created by Biesse Group, an Italian multinational company operating in the market of machines and systems for the processing of wood, glass, stone, advanced materials, and metal. The platform allows, thanks to the use of sensors inside the machinery, to share data and information in real-time on machinery and plants, offering the possibility to optimize productivity and maximize performance. In addition, it is possible to monitor the performance and any blockages or failures of the machines through remote control. Once the machines are identified and reported the failure, it will contact directly the Biesse control center, which in turn will send a technician for repair. On the other hand, the platform is also composed of an online portal that allows customers to order the parts they need, find out the price and availability, and also put machinery or spare parts up for sale.

Zhou et al. (2015) affirmed that despite the great digital revolution taking place, there is still room for improvement in the context of security, trust, reliability, traceability, and better integration of the value chain. In this regard, blockchain is proving to be one of the most effective means to aid and fulfill smart manufacturing. According to Lee et al. (2019), blockchain should be used to support the Industry 4.0 initiative and eliminate problems related to it. Blockchain would improve security and privacy by using cutting-edge cryptographic algorithms and a global consensus mechanism. The technology would also reduce bureaucracy and overheard costs thanks to smart contracts and P2P interactions. In addition, blockchain would enhance transparency by tracking components from their origin, step by step, and boost interconnectivity between devices through Master nodes (nodes with higher capacity).

The immutability characteristic of a blockchain is fundamental because every manufacturing process record can be kept, for example: which raw materials and which machines have been used to produce which products. In case of recalling certain products, because of compliance, the blockchain-based manufacturing applications can quickly isolate the products in question and pull out the related production details, thus significantly reducing the product tracking efforts.

According to Zuo (2021), the use of blockchain smart contracts combined with IoT devices will allow the machine to plan self-maintenance, self-repair, and self-update of software. In case new materials are needed, smart devices can search, negotiate, and order materials from suppliers and issue the payment directly. Any activity, related to the maintenance and repair area, conducted through a blockchain platform can be recorded, verified, and traced, which enables the resolution of any disputes by looking at blockchain transaction history. Considering that the big furniture manufacturers are enhancing their production processes based on Industry 4.0 to satisfy their customers' needs and increase their efficiency, blockchain technology can be a good solution to be implemented besides other technologies to improve traceability, security, and transparency. Therefore, the second research question regarding blockchain and the furniture industry can be devised as follows.

RQ 2. How can blockchain technology combined with a 4.0 production plant be implemented in the furniture sector?

Blockchain and Circular Economy

Environmental issues have incredibly impacted most industrial businesses' decisions. In the furniture sector, in particular, the importance of sustainability practices is requiring the reevaluation of operations management and value chain managers' actions. For instance, The WoodCircus project's main goal is to promote wood-based value chains as a key part of a circular bio-economy in Europe. One of the practices related to this project is that old wooden furniture that cannot be sold anymore and would otherwise be incinerated is gathered to give them a second life and re-circulated for local use. Upcycling old furniture into individual furniture pieces for private homes, offices, and public cultural spaces help to create awareness for local raw material collection and valorization (https://woodcircus.eu/index.php/about/). Figure 3 shows examples of making furniture from recovered woods.

Figure 3. Furniture is made by 85-100% of recovered wood (WoodCircus, 2021)

Europe continued its strongly sustainability-oriented path, consistent with the 2030 Agenda, and defined with the Green Deal at the end of 2019. The circular economy is one of the key transitions, defined as part of the 2030 Agenda and the Green Deal. Circular Economy aims to combine economic activities with environmental well-being in a sustainable way, through its cyclical wastage and cleaner production, following the three Rs principles including Reduce, Reuse, and Recycle (Murray et al., 2017).

Digitalization and Industry 4.0 technologies are considered the main enablers of the green transition toward a Circular Economy. Blockchain as one of those technologies represents a particular and avant-garde manner of sharing and storing information through its P2P system, and it is because of its unique features that many researchers believe it may have the potential to boost Circular Economy initiatives (Saberi et al., 2019).

The production of the furniture sector is extremely fragmented among independent firms, which are scattered located, and this makes it very challenging for firms to produce ethically and reduce environmental footprints. Moreover, it is particularly difficult to monitor and influence suppliers' activities when they are characterized by different environmental standards. Blockchain technology by establishing a transparent decentralized, simultaneously updated, and irreversible database among the stakeholders of networks can assign cryptographic digital identities to physical products and goods and consequently depict the provenance of a product (Bezuijen et al., 2021). This system can also facilitate reverse logistics, which is one of the central difficulties of the entire supply chain, due to poor data accessibility about goods, their location, state, and value. In this way the return of products and components (such as wood, metal, or plastics) for reusing or recycling may be much simpler, organized, and informed.

Blockchain in Circular Economy provides a technical infrastructure for CE processes since it simplifies physical and non-physical transactions of CE, optimizes CE processes, its governance structure enables transactions that do not require a trusted intermediary, and provides security improvements (Böhmecke-Schwafert et al., 2022). Another key advantage given by the use of blockchain is the achievement of high levels of coordination and cooperation. As stated by Upadhyay et al. (2021), circular economy can also benefit from simultaneous cooperation, adopting a decentralized principle of value creation and circulation as opposed to value creation and value appropriation. Applying the blockchain technology for Circular Economy processes would be helpful for recycling wood to produce panels for furniture, therefore there is no need to consume virgin wood leading to a decrease in deforestation and greenhouse gas emissions. Accordingly, our final research question is devised as follows:

RQ 3. Can blockchain technology developed in the furniture sector enhance Circular Economy practices?

To the best of our knowledge, there are no published works that deal with the subject of blockchain technology in the domain of supply chain management in the kitchen furniture sector. The technology is still in its early stages, nevertheless, there is plenty of available literature concerning the application of blockchain technology in other sectors, which will provide us a broad overview of the state of art. To see whether the assumptions that have characterized other industries were also evident in the furniture industry, three research questions were first proposed and analyzed through a theoretical approach and then applied to the business case.

The first question is "How can blockchain technology guarantee information transparency in the furniture sector?" and it has been addressed by describing the principal characteristics of the sector and highlighting the critical points that can be solved through the adoption of technology. The second question is "How can a blockchain-based traceability system combined with a 4.0 production plant be implemented in the furniture sector?". The answer to this question was given by listing the main technologies inherent in Industry 4.0 which combined with Blockchain technology could improve the competitiveness of the sector. The third question is "Can blockchain technology developed in the furniture sector enhance Circular Economy practices?" and it is focused on how most of the barriers the industry needs to overcome in order to achieve a certain level of circular economy can be easily lowered by the implementation of blockchain technology.

In the next section the importance of the furniture industry in Italy is described and our case study Cubo Design S.r.l is introduced. The company analyzed as a case study belongs to the Kitchen & Dining segment. After a comprehensive introduction about the company, the goal is to answer research questions by applying theory to the reality of the company. Finally, at the end of the chapter, the challenges, conclusions, and managerial suggestions are provided.

FURNITURE INDUSTRY IN ITALY AND CUBO DESIGN S.R.L

To date Italy is fifth in the world in terms of trade balance with a value of $7.2 billion ($8 billion in 2021), preceded by China (99.8 billion), Vietnam (12.4 billion), Poland (11.8 billion), and Canada (7.3 billion). In exports, it ranks third in Europe after Germany and Poland with exports of $14.6 billion. In Italy the furniture sector has a strong historical connotation, it was born in the second half of the twentieth century and contributes greatly to the development of the national economy, thanks to the dynamism, creativity, and flexibility of the companies and, above all, the solid territorial relationships that have characterized the development of this sector. The Furniture sector represents one of the so-called "4 A" of Italian manufacturing, a symbol of "made in Italy" globally along with the Clothing, Automation, and Food sectors. Over the past decade, the sector has suffered severely from the various crises that have swept through the country, starting from the drastic downturn in the domestic market following the 2008 crisis, which led to a sharp reduction in enterprises and employment in the sector.

In 2015 the sector went through another rough stage in its history, fronting both huge opportunities and massive risks. On one hand, Italian furniture companies held great potential related to exports, thanks to the strong attractiveness of "made in Italy" products, and on the other hand, were discouraged by the risks related to a domestic market in a deep recession that was unable to sustain growth rates adequate for the sector. Finally, 2021 represented an important year in consolidating the recovery of the sector and for Italy after the pandemic, nevertheless, the long-awaited recovery could be thwarted due to war conflicts in Ukraine. With a transparent supply network based on high-tech principles, Italy would be more competitive and stand out not only for its unrivaled Made in Italy but also for the transparency of its sustainable practices. Blockchain despite its degree of novelty, combined with 4.0 technologies, could really be the turning point for the Italian furniture industry, taking it to the top and becoming the leader among global furniture exporters.

Our exploratory investigation into the potential use of Blockchain technology in the wood furniture sector leads us to analyze a case study in this sector. The case selected for this work is "Cubo Design S.r.l" since it can be considered one of the most innovative and cutting-edge companies in the furniture sector, which has made digital transformation its real strength. The company, founded in 2005 by Giuseppe and Antonio, respectively father and son, is specialized in the manufacture of furniture parts and accessories, producing, and distributing kitchens worldwide. Located in Abruzzo -specifically in the province of Teramo- the company could enjoy a great advantage due to its fifty years of experience in the furniture sector.

The founder of the family group, Giuseppe, was born as a wood craftsman and passed on his art and passion as a carpenter to his son Antonio, conveying to him that wealth of knowledge brimming with expertise and mastery of wood. This huge background of experience and proficiency allowed Antonio to fully express himself as an entrepreneur in the field of furniture and his creative know-how, which has in turn allowed Cubo Design to conquer in a few years both domestic and foreign markets, selling kitchens all over the world under different proprietary brands.

Binova and Miton are the most iconic ones: the former, an Italian historical brand that has been acquired by Cubo Design from an Umbrian company, belongs to the "Premium" segment, characterized by a high price range, where materials and workmanship are distinguished by quality, design, functions of use and style of innovation; the latter belongs to the "Contract Entry level" segment in the medium price range, which offers solutions with refined design, respectable ergonomics, and durability. This diversification has enabled Cubo Design to meet every market demand thanks, above all, to the wide choice of finishes, colors, and materials such as melamine, veneer, matte, and glossy lacquers (to name a few). The quality of the product represents an important potential for the company to meet the national and international demand for "Made in Italy".

Currently, Cubo Design is not using Blockchain technology, but the company's ability to take on a major challenge such as digitalization and to change the paradigm of craftsmanship allowed us to envision the adoption of

Figure 4. Sample kitchen designs from the case study (source: Cubo Design official website)

further innovations. Cubo Design's willingness is to strengthen and improve business relationships with its suppliers by creating a transparent and always-informed supply chain and Blockchain technology can be a good solution for achieving these goals. Several meetings have been conducted with Cubo Design's CEO and employers to collect data on the firm activities, vision, mission, and philosophy.

CUBO DESIGN S.R.L

Blockchain and Traceability

Hereby we discuss RQ1 (How can blockchain technology guarantee information transparency in Cubo Design?) to explain how Cubo Design could advance information transparency within its supply chain by introducing blockchain technology. However, to fully understand the potential effectiveness of blockchain technology in this precise case, it is fundamental to study the company's supply chain and difficulties in recent years because of the pandemic and other crises that adversely affected its efficiency.

Cubo Design company operates in more than 50 countries, especially in France, Portugal, China, and the United States. The year 2020 ended with foreign market sales accounting for 62,8% of total sales, revealing the company's strong inclination and presence towards the global market. In

2021 this percentage has been declining and currently stands at 55,34%, demonstrating the impact of Covid-19 on foreign transportation costs and the company's willingness to balance the scales of the two markets.

Cubo Design kitchens are designed with three simple components: furniture, worktop, and household appliance. For what concerns the production of a piece of furniture, the raw materials include wood, steel, aluminum, glass, and plastics. The most important element is wood, such as Chipboard or Medium-density fiberboard (MDF), which arrives at the factory in the form of panels. The procurement is carried out through semi-annual programs where the established delivery is made daily and amounts to about 160 panels, with one-day stock. In the panel processing department, three major stages of the production cycle are carried out: sectioning, squaring, and edging of the semi-finished products. Hence, panels' usage is to create semi-finished products which can have two different natures: reference semi-finished products which are most affected by the customization of the product and therefore, they are unique or otherwise at very small quantities; high-turnover semi-finished products, that are those furniture components that are least affected by product customization and are therefore produced in large quantities. An important activity to be done in order to personalize the semi-finished product is varnishing, which is carried out in-house with the utmost respect for the environment. Following the cleaning activity, the semi-finished products are ready to be assembled and arrive at the use station fully automatically and in the established sequence.

Steel and aluminum are used as decorative and assembling components, while the use of plastic occurs in very small quantities, as even packaging is done with paper. Worktops, as the name suggests, are designed with tops that are generally made of porcelain stoneware or marble made with digital printing, an activity in which Italy is considered one of the leading players. As we can notice, even though the company, thanks to its new 4.0 plant, is incredibly doing much of the processing necessary to make the product in-house but the raw materials remain as items that the company has to source.

As it was mentioned before, the furniture sector can have a complex supply chain. It is extremely fragmented among several suppliers located in every part of the globe. Generally, the company and the sector are still facing difficulties in production because of delays, shortages of raw materials, closed production facilities, transportation blockages, and rising prices. The CEO of Cubo Design highlighted other gaps within the supply chain, concerning the fulfillment of sales conditions in terms of reliability, the observance of a quality standard by both parties involved (supplier-customer), and regulatory conformity to some materials.

Due to the spread of Covid-19 and the following world economic crises, in the last two years, the entire furniture supply chain has experienced notable delays and breakages. During the pandemic, Cubo Design faced a 5% drop in exports on sales, because of the imposed shutdown of the production plant and the stoppage in transportation with lockdowns. The consumers reevaluated their purchasing priorities, focusing on the place they experienced the most during the various lockdowns, namely the home, and there has been an increasing focus on quality and the environment, for these reasons the post-closure period brought a 20% increase in sales for the company.

Nevertheless, the Russia-Ukraine war has contributed to exacerbating the disruption already underway, implying the rise of transportation costs, the lengthening of delivery time, the increase in raw material costs, and the obstruction of import and export. In January 2022, the company witnessed an unprecedented shock in the European energy markets, short-term gas prices were "five times higher than their 2021 average. Aluminum price has reached 3,849 U.S. dollars per metric ton, a historical peak since 2008 (Statista.com). Prices are increasing rapidly, but also production continues to decline, partly as a result of the closure of many smelters, especially in Europe, where the cost of energy for many producers is now unsustainable.

The company is facing an increase in raw material prices from a minimum of 15% and a maximum of 25%. Electricity and gas have increased by 735%. Other problems involved are delays in the sourcing of two materials in particular: urea, a natural glue essential for some specific processes, and kaolin, a material widely used in porcelain stoneware. The CEO also added, "We have shifted from production disruption to raw material shortages. A series of events have converged to define what is called the perfect storm".

The downstream solution to all these matters is the improvement of the information system that connects all participants in the supply chain. As in many other sectors, data are of primary importance and in the Cubo Design supply chain, they are obtained from different sources, they are unstructured, and have different formats. That is precisely why blockchain, through its ability to process big data, could unify information and make it available to the entire supply chain. Blockchain technology could bring several distinct advantages to the furniture supply chain, such as a permissioned distributed ledger.

In the case of Cubo Design, the technology will foster the creation of an exclusive transparent, and trusted economic environment, in which all the participants of the chain, from suppliers to clients and retailers, are informed about all the stages that the product has gone through and its current state. All supply chain members from upstream to downstream can access precise and real-time information about the products. A digital twin is made for every physical product, to which each production process is linked and recorded in a distributed database. Each product's "history" is traced together with its values of authenticity and uniqueness, also certifying its sustainability in terms of the environmental and social footprint of business activities. The information is reachable through various marking systems, which are also characterized by the usual incorruptibility criteria inherent in blockchain technology. In addition, a more informed system could ensure a timelier and more responsive organization so as to increase production in stressed regimes. Figure 5 describes the information flow before and after implementing blockchain technology.

To have a clear picture of how blockchain could be introduced, consider Cubo Design's procurement of porcelain stoneware. The supply chain is composed of two actors: company A and company B. The former produces clay and kaolin, and supplies company B; the latter, in turn, processes the two materials, creates porcelain stoneware, and supplies Cubo Design. With the introduction of blockchain, the company would be able to cope with the emergency both on the supplier and customer side. On one hand, the company will be informed about all the possible breakages of company A, without the intermediation of company B (as it usually is), because real-time data are shared among the blockchain platform. On the other hand, all consumers who have placed an order with the stoneware worktop can receive all the news regarding the status of the raw materials and the final product. Therefore, this system, not only will deliver greater transparency but will also reduce mediation costs, and misunderstandings, and would provide a cohesive view of the chain to the end customer.

Other benefits linked to the implementation of blockchain for Cubo Design may concern trade facilitation through smart contracts, which would ensure the security of each player's adherence to the terms of the contract since the information exchanged is encrypted and would also encourage the use of automated processes, reducing human error and increasing accuracy and precision. Many consultancy firms are working to provide solutions for the implementation of blockchain technology in the field of furniture manufacturing, according to EY, blockchain is the ideal technological solution for transparently manifesting and certifying intellectual innovations, such as designs, industrial models, and patents; in the furniture sector blockchain

Figure 5. Information flow before and after Blockchain

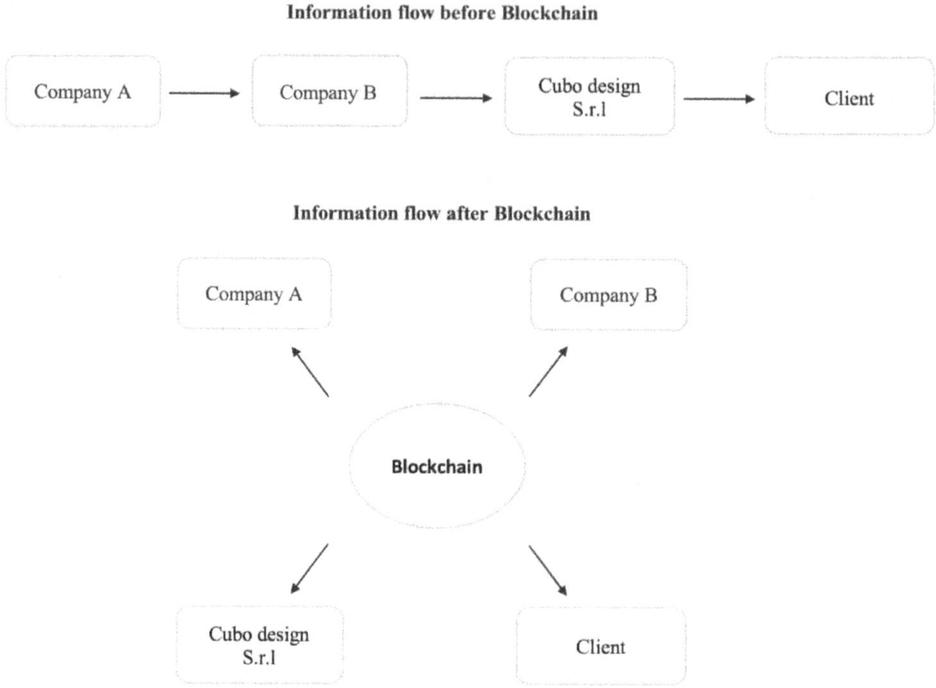

works as an accelerator of culture and industrial value but also of efficiency and reduction of costs associated with protecting innovations.

Blockchain and the 4.0 Production Plant

This section aims to answer RQ2 (how can blockchain technology combined with a 4.0 production plant be implemented in the furniture sector?), through the development of a solution that combines the 4.0 technologies present within the Cubo Design plant with blockchain technology.

In Cubo Design, over the years, as sales and orders increased, the first production problems emerged, and operating costs began to rise. The main stages of the company's system were product conception, design, and production according to a traditional work-island scheme: picking semi-finished components from the warehouse; customizing the product according to consumer needs; final assembly; and shipping. Many of the external components required for the production of the finished furniture, including accessories such as hinges and drawers, were supplied by third parties.

This business model required a great deal of business and logistical organization, making it fundamental to change the entire production apparatus and invest in machinery and employees. In addition, there were limitations to increasing the factory area. Therefore, the board of directors made a decision to build a new plant that would give adequate answers and solutions for the coming years. The idea of a new production plant came up in 2016, after a brilliant intuition of the CEO Antonio who, inspired by car factories, imagined an integrated production system for Cubo Design that would cover all stages of the production process in a single plant, revolutionizing the concept of a furniture factory, enhancing efficiency and productivity without stiffening the organizational structure too much.

According to the CEO "Until a few years ago, kitchen factories were thought of as "screwdriver factories" (a term which usually refers to Honda's factories), which identifies factories that have only final assembly lines. This concept enabled suppliers to raise prices at will, making it very challenging to develop a competitive cost strategy. Our plant has shifted from such a production system to a verticalized one, where the entire process is carried out in-house, lowering both transposition costs and material stocking costs. Therefore, the company has moved from mass production (which is commonly used by almost all Italian kitchen manufacturers) to batch 1 production". In 2018 a radical reorganization of its main assets has been implemented in which efficiency, effectiveness, and respect for the environment are concretized thanks to fully integrated technical, IT and energy solutions: FactoryOne is born (Cubo Design Official Website).

The innovative project is based on the application of all the latest technologies for processing wood and its derivatives, designed for just-in-time production. In recent years, manufacturers of woodworking machinery have inherited innovative performing techniques, already used in other sectors, and their implementation has allowed the realization of plants suitable for the production of small quantities of product, very customized, on continuous flow cycles. As the "National Industry 4.0 Plan" came into effect, the goal for Cubo Design was clear: seize the opportunity offered by the Italian government and bring the entire production cycle back house, so as to achieve savings based on economies of scale, maximum product customization and lower production cost. The new project, which still represents one of the first examples of a "Smart Factory" in Italy, is concerned with the realization of an integrated production system that allows the combination of product

Figure 6. FACTORY-ONE (Cubo Design's official website)

customization, process standardization, and real-time production planning, with the dual result of shortening delivery times and achieving economic, energy and raw material savings and efficiency.

The new factory involves the use of production lines with a high level of automation, supported by a series of digital software applications to manage all business processes: from product configuration (designing items in a series of variants to give rise to all the final product combinations required by customers) to scheduling the machines in the factory; from planning supplies (panels, semi-finished products, hardware, and other raw materials, and marketed products) to setting up shipping plans: a "Smart Factory" in all respects. Figure 6 indicates some processes in the new factory. Daily production is around 750-800 pieces of furniture i.e. 55-60 kitchens and is divided into 5 departments: panel processing department, processing-assembly-packaging department, painting and lacquering department, and the warehouse-shipping department, energy-compressed air-thermal systems department.

The introduction of 4.0 technologies in the new Cubo Design factory was a turning point for the Abruzzo-based company, which can thus benefit from the potential related to the digitization and automation of processes. Now the production cycle includes all the main processing stages, which, starting from the raw material, lead to the creation of the finished furniture. Each production process is interconnected through the use of new 4.0 technologies so that all departments of the plant form an organized and efficient enterprise system. As a result of adhering to "Industry 4.0," order handling and delivery times have been significantly shortened, and inventory references have also decreased by 80 percent. Furniture currently stays in the storage department for no more than 24 hours.

Starting from the product launch, through the implementation of the digital product configurator, each user is able to understand and configure the desired product. After the customer order, the digital configurator generates all the documentation for both the commercial aspects (order confirmation, packing list, specific drawings) and the production aspects (technical drawings cam files, labels). The information system is complemented by a supervisor (M.E.S.) who, by dialoguing with the control systems of each plant, enables the distribution of the information generated by the product configurator and the management system to the various recipients inside and outside the company: workstations, warehouses, and suppliers. The innovative technologies introduced in each plant allow the equipment in one department to be combined with the equipment in the next department, making it possible to achieve a fully integrated production cycle capable of increasing both hourly output and production flexibility, shortening production lead-time from the previous 6 to the current 2 weeks, and considerably reducing production costs. Figure 7 shows a diagram that summarizes all the production processes carried out within the company.

In the recently built factory, the property is no longer the "mere collector of third-party production activities, but they guarantee its product, as they monitor and know in detail all the steps of industrial processing". The categorical and systematic control of each stage of the manufacturing process thus certifies full consumer care, which has always been a priority of Cubo Design. What is new is that this innovative transformation of the production method not only projects into the future but seems to anchor itself in the past, in which high quality depended exclusively on the skill of craftsmen. These artists were capable of shaping their works, devoting themselves to the study and refinement of every single aspect of their creation with the same passion employed today, inside the factory, by the entrepreneur and the R&D team in the choice of materials and the attention to detail.

The manufacturing process is accomplished through a path of a linear trend, consisting of highly technological, interconnected machinery marked by absolute precision, constantly guaranteed through specific controls carried out by skilled labor. The company has installed RFID sensors that, exploiting the connection to the network, make it possible to monitor the various stages of industrial processing and to know in real-time the status of the equipment on which they are integrated. It thus becomes easier to take prompt action on machine breakdown, also thanks to the immediate reporting of failures or malfunctions, accomplished by the equipment itself via digital software, to the respective service centers. Through cloud computing technology, it is also possible to collect countless data which are stored in total security and used to continuously improve the production flow.

Nevertheless, in Cubo Design RFID and cloud data are implemented just for the in-house processes, thus not involving the entire supply chain. With the introduction of blockchain, Cubo Design could make RFID data available throughout the supply chain, accelerating its effect and increasing transparency. The introduction of blockchain would not mean the replacement of RFID technology, but a perfect combination of these two technologies. The new solution proposed includes the use of existing RFID data and EDI (Electronic Data Interchange) connections, creating a neutral platform, where data can be pulled together through Blockchain to get more transparency.

The integration of the systems will allow the company and the entire supply chain to have a more informed system. The cooperation of blockchain and RFID technologies could thus lead to greater transparency between the company and its suppliers, but it will also improve reverse logistics for product ineptitudes. Cubo Design has a team whose work is dedicated to the analysis and resolution of all consumer complaints, however, with the implementation of blockchain technology, it can assist in data sharing, record every transaction, and consequently trace the error such as inaccurate invoicing, late shipping, or product defects.

In addition, RFID technologies still use a centralized manufacturing network and a "third-party trust operation". With the adoption of blockchain, Cubo Design could operate in a decentralized way, without the necessity of an authentication system, having the same volume of reliability.

Figure 8 shows RFID and blockchain integration in Cubo Design. This model is considered for the company to take the most advantage of integrating its current system which depends on RFID technology with blockchain. Any necessary information is collected by RFID sensors at each stage of production automatically and is stored on the blockchain. Customers can trace the final product history and the company can trace every process from the initial material storage to the last mile delivery. By developing this technology for the suppliers, the efficiency of the supply chain would increase by enhancing transparency between the company and the suppliers, improving reverse logistics, and implementing blockchain-based smart contracts.

Figure 7. Cubo Design's main production flow

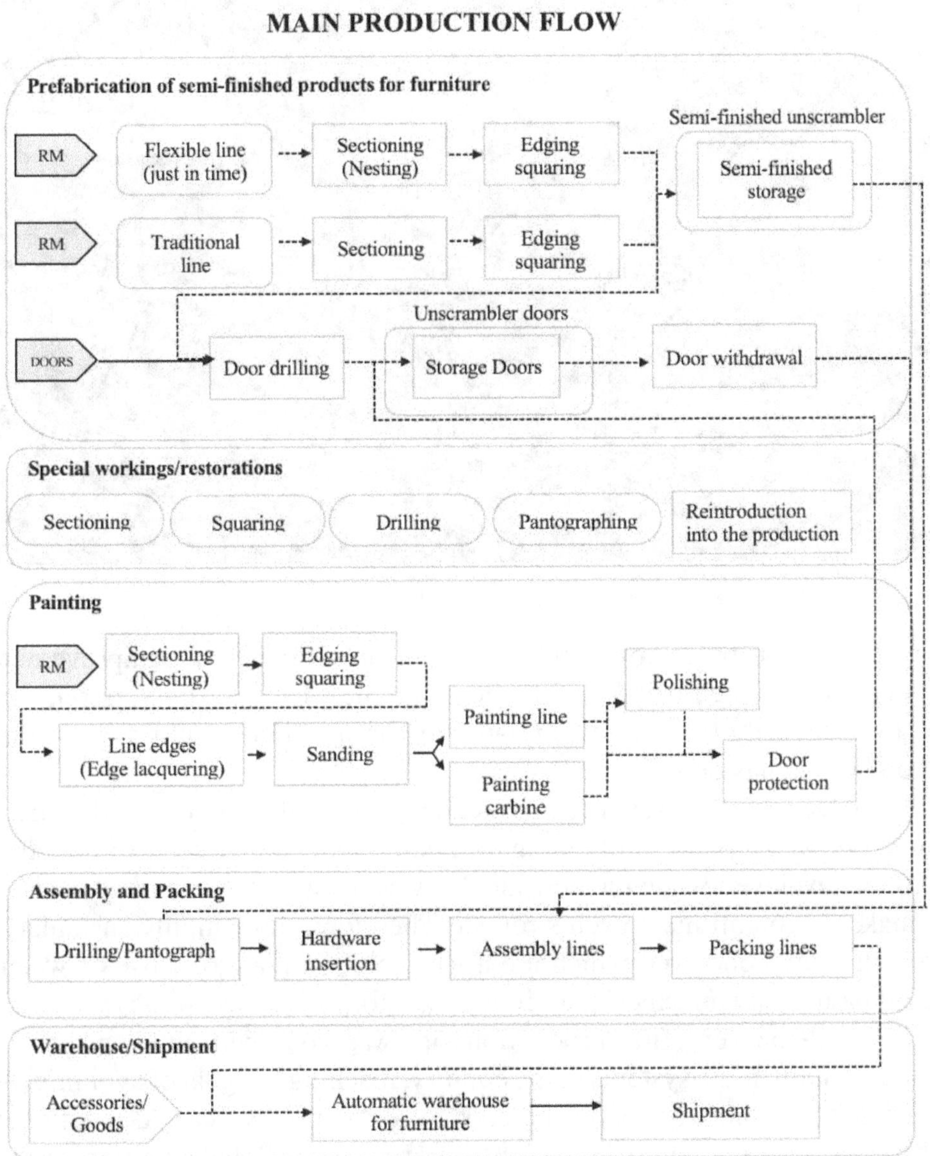

Blockchain and Circular Economy Practices

The goal of this section is to discuss RQ3 (can blockchain technology developed in the furniture sector enhance Circular Economy practices?) in a

Figure 8. RFID and blockchain integration in Cubo Design

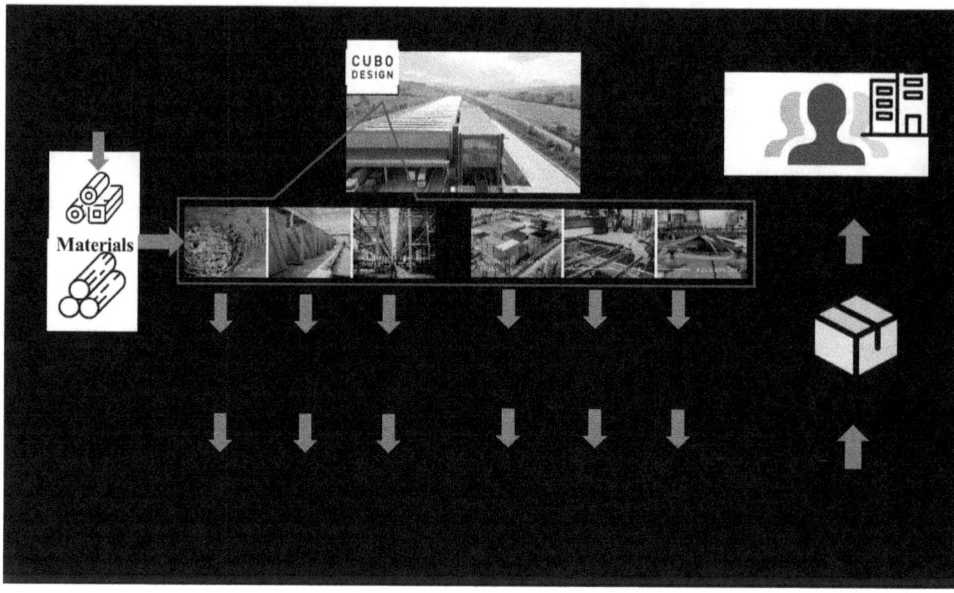

way that demonstrates how blockchain could contribute to the improvement and creation of a certified circular economy network. For this reason, first we need to study the activities related to Circular Economy practices that Cubo Design has done in recent years.

Cubo Design has always been particularly involved and concerned with the protection of the planet. Following the adoption of an innovative production method developed when the new factory was opened, the company decided to make its commitment even stronger. Therefore, three highly sustainable projects were conceived which are a trigeneration plant, the use of water-based paints, and an experiment in green mobility.

In 2012 the very first production site was equipped with one of the earliest photovoltaic systems in the country, reaching the goal of avoiding the consumption of 600 tons of oil and 2000 tons of Co2 each year. As mentioned before, in 2018 the company took up an exciting new challenge: opening a factory with a 25,000-square-meter facility, which the founder called Factory-One. Today the entire production site is powered by a trigeneration plant capable of generating, from methane, 3 types of energy: electric, thermal, and refrigerated. In this way, the company succeeded in producing simultaneously: warm water, used for space heating and other business processes; cold water, which usage is made for varnishing operation and conditioning during the

Figure 9. Nonpulluting energy that powers facroyr one (source: Cubo Design official website)

summer and electricity, which is fundamental to the proper execution of all business processes, in offices and within production.

The reason why Cubo Design chose to install and adopt a trigeneration plant concerns energy wastage. In plants where electricity and thermal energy are produced separately, there is a useless heat loss that can't be reused in other ways, so it is spread in the surrounding environment, causing 70% of energy wastage. Thanks to the trigeneration plant, the company losses of energy are minimized, since the waste heat is reused for company production flows, to warm and cool, not only the factory itself but also the water used in the site. Nonetheless, the new plant ensures a significant reduction in methane consumption and prevents the spread of an abundant amount of polluting substances, thus protecting, and safeguarding the quality of the air in the territory. In addition, the compressed air, energy, and vacuum systems are all equipped with sensors and innovative technologies, so that the management information system can control and monitor, even remotely, the plant's efficiency and energy consumption.

In 2018, one more change transformed the company's attitude toward the environment: the usage of water-based paints. The head of manufacturing explained that the adoption of these new paints was dictated by two priorities "the wellness and the satisfaction of our workers and of people who choose our products" (Cubo Design official website). The particularity of these paints

Figure 10. Cubo Design presents A+ class water paints (source: Cubo Design official website)

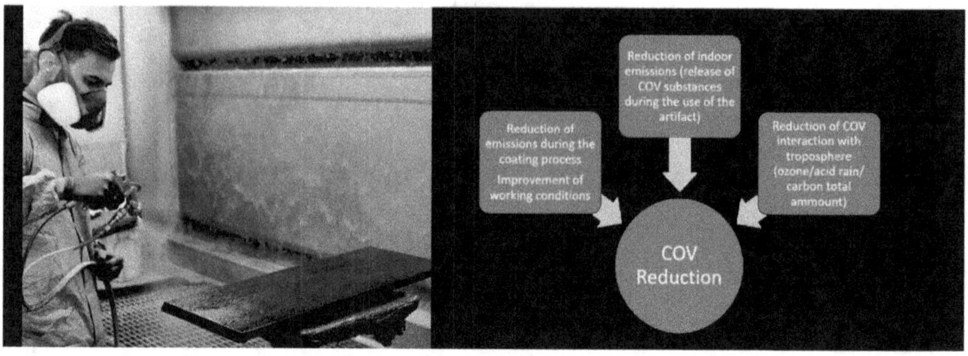

is that water is used as a diluting agent, to which is added a small amount of solvent, around 2-3%, while normal coats have a concentration of 30-90% of solvent. The adoption of water paints has significantly decreased emissions (90% VOC reduction) thanks to a very reduced number of harmful substances, which have been substituted with water vapor.

In terms of performance, thanks to these new paints, the company can guarantee a greater level of color stability of the product. Traditional coatings have the inclination to yellow after light exposure, through water paints instead the color does not change over time. This path has been made possible through collaboration with an Italian chemical company, with which Cube design shares the same values in terms of innovation and environmental protection.

Another component of the company's Corporate Social Responsibility is about to be added with the introduction of the project "Easy mobility". The launch of the project is planned to revolutionize the company's employees' movements, through new suburban mobility. Workers will be equipped and encouraged to use electric cars provided by the company. The energy needed to run the machine will also be provided by the company through recharging stations made with the electricity produced in excess from the plant. According to the CEO "The benefits will be immediately evident: we will make our workers' life better without pollution".

As mentioned, Cubo Design is very active in developing projects under the banner of environmental protection. In addition, the company is involved in a very important process for Italian land: wood recycling management. Millions of tons of wood waste are thrown away every year and there is very little awareness and knowledge regarding the issue of wood disposal. When

Figure 11. The project "Easy mobility" (Cubo Design's official website)

wood is landfilled, not only takes away space from non-recyclable waste but also releases methane and if is not captured and used (very few facilities are equipped to do so) it is inevitably released into the air.

Italy, as already stated, retains the gold medal for Circular Economy among the European Union's leading economies (CEN, 2021). In the five years 2016-2020, the number of enterprises that invested in sustainability and efficiency grew by over 441,000 (31.9 percent of the total), that is, 1 in 4 enterprises, values that have increased compared to the immediately preceding period 2015-2019. In the manufacturing industry, 1 in 3 enterprises made eco-investments in the 2016-2020 period (data have been taken from www.federlegno.it). In addition, the tons of wood collected and sent for recycling reached 1.8 million, with an increase in volumes of 7.8 percent over the previous year and a 64.75 percent share in the recycling of wood packaging. Thus, the target set by the European Union of 30 percent recycling by 2030 was doubled (data have been taken from rilegno.it).

The first sustainable step in this context for Cubo Design was the decision to use recycled wood, such as chipboard panels. The second step has been the choice of a sustainable partner, which shares the same vision and proudly carries it forward. The supplier is the largest wood waste transformer in the world and collects annually 1.5 million tons of used wood. Located in Italy, the vendor has been operating under the circular economy philosophy for 30 years. The company production model is based on the recovery of post-consumer wood, the collection of wood waste through a network of centers, the processing of waste, and the production of chipboard panels. Since 1995, the firm's ecological panel has been the first to obtain the FSC® Recycled 100% and FSC 99% certifications (melamine-faced, fire-retardant, and water-resistant).

Cubo Design in this Circular economy network contributes to the collection process, as all the waste obtained from the processing and cutting of the chipboard panels is vacuumed and collected, then placed inside a container and shipped to the nearest collection center ready to be brought back to the supplier.

However, Cubo Design and its suppliers could improve their support and commitment to the circular economy through the adoption of blockchain technology. In this regard, the role of blockchain could facilitate reverse logistics activities, since they require a huge effort in product data availability. The transparency given by the uncorrupted systems of blocks allows stakeholders to obtain focused monitoring and control abilities as materials' history can be traced in the whole product life cycle. Blockchain technology could also provide a platform for all those companies who want to exchange and trade their wastes and reestablish value. Companies could cooperate directly without any intermediation and therefore increase profit margins. Through the help of electronic sensors and tracking devices, blockchain can acquire the location and state of waste and make data available on the shared platform.

Moreover, some of Cubo Design's suppliers are located geographically distant and belong to nations with different legislations from Italian ones, therefore it is difficult to monitor their activities and certify that all best practices are being implemented. Through the adoption of blockchain, Cubo Design could create an informed chain where all the members can freely evaluate the green activities of each producer. In addition, the future trends of the market are becoming more and more projected on sustainable projects, to the extent that especially millennials and Gen Z will choose and buy sustainable furniture. For this reason, it is necessary for firms to have a system that allows them to demonstrate in a transparent and verifiable way their commitment to environmental issues. Having a certified circular economy network could therefore lead to greater attractiveness to consumers and increasingly reinforce the value that Made in Italy already possesses.

Figure 12. Implementing blockchain technology in Cubo Design considering circular economy aspect

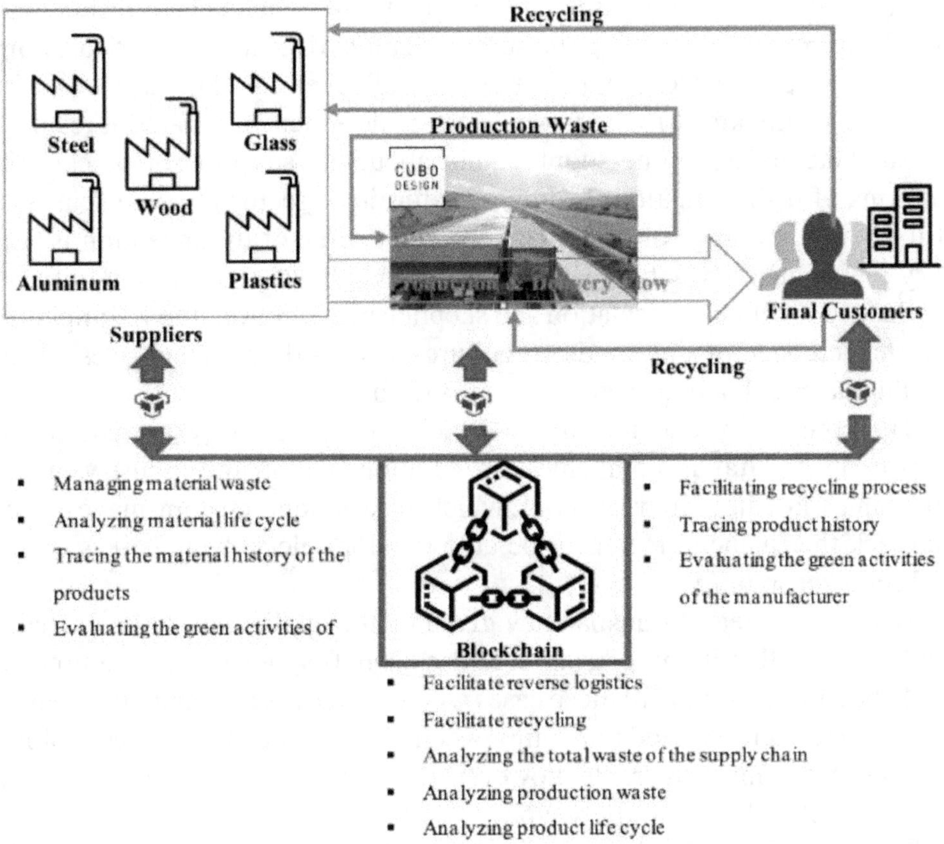

Figure 12 indicates a sample conceptual model for implementing blockchain technology in Cubo Design considering the circular economy aspect. Blockchain helps the proposed company to improve its ability in traceability to evaluate the production waste, material waste, recycling ability, energy consumption, reverse logistics, and green activities in whole steps of the supply chain. The company, suppliers, and customers can trace the final product life cycle and the material history with a high level of security. Transparent information flow enhances the ability to devise short-term and long-term strategies for achieving circular economy goals.

Blockchain Challenges

Although the research focuses on the potential that blockchain technology could have in revolutionizing the supply chain management of the furniture sector, it is appropriate to highlight what challenges could be faced.

The first limitation to be addressed is the achievement of a shared agreement by all stakeholders in the supply chain about the adoption of blockchain in terms of disinformation. In Italy, the furniture sector in some senses is still in the early stage of digitalization. The degree of digitalization is still very low, which not only creates a low attitude toward innovation but also produces a lot of misinformation and skepticism. Moreover, most companies are characterized by old mindsets, cultures, and work methodologies which would make such a major change very difficult.

The complexity and novelty of Blockchain technology are additional constraints for Italian companies. In addition, within the industry, many companies are true craft workshops, which make unique, custom-made pieces and lack the financial means to be able to adopt blockchain as it is costly and time-consuming.

With the degree of transparency given by the application of blockchain, we have seen that it could bring countless benefits, however, sometimes it could be a limitation for all those cases where privacy is fundamental. Hence, privacy and security need to be improved and guaranteed as the technology is still nascent and vulnerable, however it can be considered a general issue not specifically linked to the sector.

Therefore, it can be deduced that many companies in the sector would lack organizational readiness, digital infrastructure, expertise, and financial resources, which could be all solved by ad hoc incentives and training programs.

CONCLUSIONS

In today's context of transformation and innovation, equipping industries and sectors with the new technologies to modernize factories and production processes is crucial for companies to maintain high business competitiveness. This chapter aims at discussing the potential operational, and environmental benefits that firms in the furniture sector could obtain with the application of blockchain technology in their supply chains.

Indeed, blockchain has the capacity to considerably improve reliability and transparency in the whole steps of the supply chain, from tracking materials and production waste to the reduction of processing times and enhancing reverse logistics. Blockchain technology has the ability to improve the efficiency of supply chains with complex networks, thanks to its instantaneously updated and immutable database shared among the players of the chain.

Managers should evaluate the application of blockchain in their SC in cases in which transparency and security are missing elements. In an environment where trust is lacking on both sides and therefore an adversarial relationship is being established, the introduction of smart contracts could protect both parties from unfair practices, eliminating all those partners who are unwilling to act for the benefit of the industry but only for profit.

In addition, managers should also consider that smart contracts have countless potential to set certain parameters that would automate many processes such as pricing and quantity definition. The adoption of blockchain technology necessitates that all the actors of the SC share the same willingness and understand the complexity, benefits, risks, and implications linked to it. It is almost mandatory to entrust specialists in the field (whether they are consulting firms or just experts) who will follow the project step by step, shaping it according to the needs.

In this study, Cubo Design S.r.l. is chosen as the case study for its brilliant attitude towards innovation. In fact, Cubo Design is not currently using blockchain technology, but the company's ability to take on a major challenge such as digitalization and to change the paradigm of craftsmanship allowed us to envision the adoption of further innovations through the interviews conducted with key people within the company. The company is extensively described in all its facets, including the paradigm shift toward automated production and the company's longstanding commitment to the environment.

In Cubo Design, blockchain technology would foster the creation of an exclusive transparent, and trusted economic environment, in which all the participants of the chain, from suppliers to clients and retailers, are informed about all the stages that the product has gone through and its current state. The new proposed solution includes the use of existing RFID data and EDI (Electronic Data Interchange) connections, creating a neutral platform, where data can be pulled together through blockchain to get more transparency.

Blockchain could also help Cubo Design and its suppliers to boost their commitment to the circular economy, thanks to its uncorrupted systems of blocks where materials' history can be traced in the whole product life cycle. It helps to create an informed chain where all the members can freely evaluate the green activities of each producer which leads to greater attractiveness to consumers. In other words, it gives the ability to different players in the supply chain to analyze the total waste of the supply chain and production waste as well as product life cycle and material life cycle. It facilitates the recycling process, managing material waste, tracing product and material history, and reverse logistics. The manufacturer and all the suppliers would be evaluated by the customers in terms of green activities such as reducing energy, deforestation, and chemical materials and their ability to provide spare parts, proper designs for repairing products, and recycling strategies.

Some of the critical issues we have just defined represent the starting point to properly understanding the potential and the benefits that the introduction of such an innovative system as blockchain technology could bring. As a result of the analysis conducted, blockchain technology can be considered one of the greatest innovations of our time in terms of supply chain management and circular economy, keeping in mind that, to date, it has not yet shown its full innovative potential. Although blockchain is still in its early stages, if combined with proper technologies, could really mean a big breakthrough in terms of efficiency and competitiveness. Therefore, this chapter represents the starting point for future research, which will employ more quantitative and engineering data and methods of evaluation.

ACKNOWLEDGMENT

This contribution is a part of a project that has received funding from the Lazio region ESF operational program 2014-2020, Axis III- Education and Training, Investment Priority 10.ii) - Specific Objective 10.5, Pivotal Action 21 – Call for proposals "Contributi per la permanenza nel mondo accademico delle eccellenze" – Project Title: "Implementazione di progetti di digitalizzazione responsabile nelle supply chain" - CUP (Unified Project Code): F86J20002710009.

REFERENCES

Agrawal, T. K., Kumar, V., Pal, R., Wang, L., & Chen, Y. (2021). Blockchain-based framework for supply chain traceability: A case example of textile and clothing industry. *Computers & Industrial Engineering*, *154*(4), 1–12. doi:10.1016/j.cie.2021.107130

Agrawal, T. K., Sharma, A., & Kumar, V. (2018). Blockchain-based secured traceability system for textile and clothing supply chain. In *Artificial intelligence for fashion industry in the big data era* (pp. 197–208). Springer. doi:10.1007/978-981-13-0080-6_10

Appelhanz, S. (2013). Tracking & Tracing-systems in the Wood Supply Chain: Opportunities and Challenges. *AMCIS Proceedings of the Nineteenth Americas Conference on Information Systems.*

Baygin, N., Baygin, M., & Karaköse, M. (2020). Blockchain Application in Mass Customization: A Furniture Sector Example. *2020 International Conference on Data Analytics for Business and Industry: Way Towards a Sustainable Economy (ICDABI)*, 1-5. 10.1109/ICDABI51230.2020.9325630

Bezuijen, N., & Hördegen, T. (2021). *The impact of Blockchain Technology on the Transformation of the Swedish Furniture Industry towards Circular Economy*. KTH Industrial Engineering and Management Industrial Management.

Biswas, D., Jalali, H., Ansaripoor, A. H., & De Giovanni, P. (2022). Traceability vs. Sustainability in Supply Chains: The implications of Blockchain. *European Journal of Operational Research*. Advance online publication. doi:10.1016/j.ejor.2022.05.034

Böhmecke-Schwafert, M., Wehinger, M., & Teigland, R. (2022). Blockchain for the circular economy: Theorizing blockchain's role in the transition to a circular economy through an empirical investigation. *Business Strategy and the Environment*, *31*(3), 1–16. doi:10.1002/bse.3032

Cagle, M. N., Yılmaz, K., & Doğru, H. (2020). Digitalization of business functions under industry 4.0. In *Digital Business Strategies in Blockchain Ecosystems* (pp. 105–132). Springer. doi:10.1007/978-3-030-29739-8_6

Daian, G., & Ozarska, B. (2009). Wood waste management practices and strategies to increase sustainability standards in the Australian wooden furniture manufacturing sector. *Journal of Cleaner Production*, *17*(17), 1594–1602. doi:10.1016/j.jclepro.2009.07.008

de Boissieu, E., Kondrateva, G., Baudier, P., & Ammi, C. (2021). The use of blockchain in the luxury industry: Supply chains and the traceability of goods. *Journal of Enterprise Information Management*, *34*(5), 1318–1338. doi:10.1108/JEIM-11-2020-0471

De GiovanniP. (2019a). *Eco-digital supply chains through blockchains*. Available at SSRN 3488925.

De Giovanni, P. (2019b). When feature-based production capabilities challenge operations: Drivers, moderators, and performance. *International Journal of Operations & Production Management*, *40*(2), 221–242. doi:10.1108/IJOPM-04-2019-0309

De Giovanni, P. (2020a). Blockchain and smart contracts in supply chain management: A game theoretic model. *International Journal of Production Economics*, *228*(10), 1–18. doi:10.1016/j.ijpe.2020.107855

De Giovanni, P. (2020b). *Dynamic quality models and games in digital supply chains*. Springer Nature.

De Giovanni, P. (2021). Smart contracts and blockchain for supply chain quality management. In *Dynamic Quality Models and Games in Digital Supply Chains* (pp. 91–110). Springer. doi:10.1007/978-3-030-66537-1_5

De Giovanni, P. (2022). Leveraging the circular economy with a closed-loop supply chain and a reverse omnichannel using blockchain technology and incentives. *International Journal of Operations & Production Management*, *42*(7), 959–994. doi:10.1108/IJOPM-07-2021-0445

Duhaylongsod, J. B., & De Giovanni, P. (2018). The impact of innovation strategies on the relationship between supplier integration and operational performance. *International Journal of Physical Distribution & Logistics Management*, *49*(2), 56–177.

Feng, H., Wang, X., Duan, Y., Zhang, J., & Zhang, X. (2020). Applying blockchain technology to improve agri-food traceability: A review of development methods, benefits and challenges. *Journal of Cleaner Production*, *260*(7), 1–15. doi:10.1016/j.jclepro.2020.121031

Francisco, K., & Swanson, D. (2018). The supply chain has no clothes: Technology adoption of blockchain for supply chain transparency. *Logistics*, *2*(1), 2–8. doi:10.3390/logistics2010002

Furn. (2022). *Circular Economy in the furniture industry - Project 2017-1-BE01-KA202-024752*. Available at: https://www.furn360.eu/wp-content/uploads/2018/10/Circular-economy-in-the-furniture-industry-11092018.pdf

Galvez, J. F., Mejuto, J. C., & Simal-Gandara, J. (2018). Future challenges on the use of blockchain for food traceability analysis. *Trends in Analytical Chemistry*, *107*(10), 222–232. doi:10.1016/j.trac.2018.08.011

Gonczol, P., Katsikouli, P., Herskind, L., & Dragoni, N. (2020). Blockchain implementations and use cases for supply chains-a survey. *IEEE Access: Practical Innovations, Open Solutions*, *8*(1), 11856–11871. doi:10.1109/ACCESS.2020.2964880

Hader, M., Tchoffa, D., El Mhamedi, A., Ghodous, P., Dolgui, A., & Abouabdellah, A. (2022). Applying integrated Blockchain and Big Data technologies to improve supply chain traceability and information sharing in the textile sector. *Journal of Industrial Information Integration*, *28*(7), 1–20. doi:10.1016/j.jii.2022.100345

Hewa, T., Ylianttila, M., & Liyanage, M. (2021). Survey on blockchain based smart contracts: Applications, opportunities and challenges. *Journal of Network and Computer Applications*, *177*(3), 1–39. doi:10.1016/j.jnca.2020.102857

Kagermann, H., Lukas, W. D., & Wahlster, W. (2011). Industrie 4.0: Mit dem Internet der Dinge auf dem Weg zur 4. industriellen Revolution. *VDI nachrichten, 13*(1), 2-3.

Kouhizadeh, M., Zhu, Q., & Sarkis, J. (2020). Blockchain and the circular economy: Potential tensions and critical reflections from practice. *Production Planning and Control, 31*(11-12), 950–966. doi:10.1080/09537287.2019.1695925

Kshetri, N. (2018). 1 Blockchain's roles in meeting key supply chain management objectives. *International Journal of Information Management*, *39*(4), 80–89. doi:10.1016/j.ijinfomgt.2017.12.005

Lee, J., Azamfar, M., & Singh, J. (2019). A blockchain enabled Cyber-Physical System architecture for Industry 4.0 manufacturing systems. *Manufacturing Letters*, *20*(4), 34–39. doi:10.1016/j.mfglet.2019.05.003

Leng, J., Ruan, G., Jiang, P., Xu, K., Liu, Q., Zhou, X., & Liu, C. (2020). Blockchain-empowered sustainable manufacturing and product lifecycle management in industry 4.0: A survey. *Renewable & Sustainable Energy Reviews*, *132*(10), 1–22. doi:10.1016/j.rser.2020.110112

Maranesi, C., & De Giovanni, P. (2020). Modern circular economy: Corporate strategy, supply chain, and industrial symbiosis. *Sustainability*, *12*(22), 1–24. doi:10.3390u12229383 PMID:35136666

Murray, A., Skene, K., & Haynes, K. (2017). The circular economy: An interdisciplinary exploration of the concept and application in a global context. *Journal of Business Ethics*, *140*(3), 369–380. doi:10.100710551-015-2693-2

Naclerio, A. G., & De Giovanni, P. (2022). Blockchain, logistics and omnichannel for last mile and performance. *International Journal of Logistics Management*, *33*(2), 663–686. doi:10.1108/IJLM-08-2021-0415

Prencipe, M. P., Vishkaei, B. M., & De Giovanni, P. (2022). Blockchain Adoption in the Winery Industry: The Case of Cantina Placido-Volpone. In Blockchain Technology Applications in Businesses and Organizations (pp. 43-72). IGI Global.

Rejeb, A., Keogh, J. G., Zailani, S., Treiblmaier, H., & Rejeb, K. (2020). Blockchain technology in the food industry: A review of potentials, challenges and future research directions. *Logistics*, *4*(4), 1–26. doi:10.3390/logistics4040027

Risius, M., & Spohrer, K. (2017). A blockchain research framework. *Business & Information Systems Engineering*, *59*(6), 385–409. doi:10.100712599-017-0506-0

Saberi, S., Kouhizadeh, M., & Sarkis, J. (2019). Blockchains and the supply chain: Findings from a broad study of practitioners. *IEEE Engineering Management Review*, *47*(3), 95–103. doi:10.1109/EMR.2019.2928264

Schmidt, C. G., Klöckner, M., & Wagner, S. M. (2021). Blockchain for supply chain traceability: case examples for luxury goods. In *Digital Business Models in Industrial Ecosystems* (pp. 187–197). Springer. doi:10.1007/978-3-030-82003-9_12

Upadhyay, A., Mukhuty, S., Kumar, V., & Kazancoglu, Y. (2021). Blockchain technology and the circular economy: Implications for sustainability and social responsibility. *Journal of Cleaner Production*, *293*(4), 1–7. doi:10.1016/j.jclepro.2021.126130

WoodCircus. (2021). *Holy Wood.* Available at: https://woodcircus.eu/wp-content/uploads/2021/01/04-Holy-wood-WoodCircus-p.pdf

Zhou, K., Liu, T., & Zhou, L. (2015). Industry 4.0: Towards future industrial opportunities and challenges. *2015 12th International conference on fuzzy systems and knowledge discovery (FSKD)*, 2147-2152.

Zuo, Y. (2021). Making smart manufacturing smarter–a survey on blockchain technology in Industry 4.0. *Enterprise Information Systems, 15*(10), 1323–1353. doi:10.1080/17517575.2020.1856425

Compilation of References

Aazam, M., St-Hilaire, M., Lung, C. H., & Lambadaris, I. (2016, October). Cloud-based smart waste management for smart cities. In *2016 IEEE 21st International Workshop on Computer Aided Modelling and Design of Communication Links and Networks (CAMAD)* (pp. 188-193). IEEE. 10.1109/CAMAD.2016.7790356

Agrawal, T. K., Kumar, V., Pal, R., Wang, L., & Chen, Y. (2021). Blockchain-based framework for supply chain traceability: A case example of textile and clothing industry. *Computers & Industrial Engineering*, *154*(4), 1–12. doi:10.1016/j.cie.2021.107130

Agrawal, T. K., Sharma, A., & Kumar, V. (2018). Blockchain-based secured traceability system for textile and clothing supply chain. In *Artificial intelligence for fashion industry in the big data era* (pp. 197–208). Springer. doi:10.1007/978-981-13-0080-6_10

Al-Maaded, M., Madi, N. K., Kahraman, R., Hodzic, A., & Ozerkan, N. G. (2012). An overview of solid waste management and plastic recycling in Qatar. *Journal of Polymers and the Environment*, *20*(1), 186–194. doi:10.100710924-011-0332-2

Álvarez-Chávez, C. R., Edwards, S., Moure-Eraso, R., & Geiser, K. (2012). Sustainability of bio-based plastics: General comparative analysis and recommendations for improvement. *Journal of Cleaner Production*, *23*(1), 47–56. doi:10.1016/j.jclepro.2011.10.003

Andreasi Bassi, S., Boldrin, A., & Faraca, G. (2020). Extended producer responsibility: How to unlock the environmental and economic potential of plastic packaging waste? *Resources Conservation and Recycling*.

Angelidou, M. (2017). The role of smart city characteristics in the plans of fifteen cities. *Journal of Urban Technology*, *24*(4), 3–28. doi:10.1080/10630732.2017.1348880

Anthopoulos, L. G. (2015). Understanding the smart city domain: A literature review. *Transforming city governments for successful smart cities*, 9-21.

Anthopoulos, L., Janssen, M., & Weerakkody, V. (2019). A Unified Smart City Model (USCM) for smart city conceptualization and benchmarking. *Smart cities and smart spaces: Concepts, methodologies, tools, and applications*, 247-264.

Appelhanz, S. (2013). Tracking & Tracing-systems in the Wood Supply Chain: Opportunities and Challenges. *AMCIS Proceedings of the Nineteenth Americas Conference on Information Systems.*

Barros, M., & Dimla, E. (2021). From planned obsolescence to the circular economy in the smartphone industry: An evolution of strategies embodied in product features. *Proceedings of the Design Society, 1*, 1607–1616. doi:10.1017/pds.2021.422

Batagan, L. (2011). Indicators for economic and social development of future smart city. *Journal of Applied Quantitative Methods, 6*(3), 27–34.

Battaglia, M., Testa, F., Bianchi, L., Iraldo, F., & Frey, M. (2014). Corporate social responsibility and competitiveness within SMEs of the fashion industry: Evidence from Italy and France. *Sustainability, 6*(2), 872–893. doi:10.3390u6020872

Baygin, N., Baygin, M., & Karaköse, M. (2020). Blockchain Application in Mass Customization: A Furniture Sector Example. *2020 International Conference on Data Analytics for Business and Industry: Way Towards a Sustainable Economy (ICDABI)*, 1-5. 10.1109/ICDABI51230.2020.9325630

Beers, K., Schumacher, K., Migler, K., Morris, K. C., & Kneifel, J. (2022). *An Assessment of Mass Balance Accounting Methods for Polymers Workshop Report.* NIST Special Publication, 1500, 206.

Berg. (2020). *Fashion on climate.* https://www.mckinsey.com/industries/retail/our-insights/fashion-on-climate

Bezuijen, N., & Hördegen, T. (2021). *The impact of Blockchain Technology on the Transformation of the Swedish Furniture Industry towards Circular Economy.* KTH Industrial Engineering and Management Industrial Management.

Biocca, F., & Harms, C. (2002). What is social presence? In F. Gouveia & F. Biocca (Eds.), *Presence 2002 Proceedings*. University of Fernando Pessoa Press.

Bioregional. (2020). https://www.bioregional.com/news-and-opinion/fashions-failings-barriers-to-a-sustainable-clothing-industry

Bishop, G., Styles, D., & Lens, P. N. (2021). Environmental performance comparison of bioplastics and petrochemical plastics: A review of life cycle assessment (LCA) methodological decisions. *Resources, Conservation and Recycling, 168*, 105451. doi:10.1016/j.resconrec.2021.105451

Biswas, D., Jalali, H., Ansaripoor, A. H., & De Giovanni, P. (2022). Traceability vs. Sustainability in Supply Chains: The implications of Blockchain. *European Journal of Operational Research.* Advance online publication. doi:10.1016/j.ejor.2022.05.034

Bocken, N. M., De Pauw, I., Bakker, C., & Van Der Grinten, B. (2016). Product design and business model strategies for a circular economy. *Journal of Industrial and Production Engineering, 33*(5), 308-320.

Bocken, N. M., Schuit, C. S., & Kraaijenhagen, C. (2018). Experimenting with a circular business model: Lessons from eight cases. *Environmental Innovation and Societal Transitions*, *28*, 79–95. doi:10.1016/j.eist.2018.02.001

Böhmecke-Schwafert, M., Wehinger, M., & Teigland, R. (2022). Blockchain for the circular economy: Theorizing blockchain's role in the transition to a circular economy through an empirical investigation. *Business Strategy and the Environment*, *31*(3), 1–16. doi:10.1002/bse.3032

Bora, R. R., Wang, R., & You, F. (2020). Waste polypropylene plastic recycling toward climate change mitigation and circular economy: Energy, environmental, and technoeconomic perspectives. *ACS Sustainable Chemistry & Engineering*, *8*(43), 16350–16363. doi:10.1021/acssuschemeng.0c06311

Bourlakis, M., Papagiannidis, S., & Li, F. (2009). Retail spatial evolution: Paving the way from traditional to metaverse retailing. *Electronic Commerce Research*, *9*(1), 135–148. doi:10.100710660-009-9030-8

Brouwer, M. T., van Velzen, E. U. T., Augustinus, A., Soethoudt, H., De Meester, S., & Ragaert, K. (2018). Predictive model for the Dutch post-consumer plastic packaging recycling system and implications for the circular economy. *Waste Management, 71*, 62-85.

Bruzi, A. (2021). Social Entrepreneurship For Sustainable Innovations: A Case Study Of Fairphone. *Zeszyty Programu TOP*, *15*, 11.

Business of Fashion. (2021). https://www.businessoffashion.com/opinions/sustainability/a-more-circular-fashion-industry-will-require-a-collective-effort

Cagle, M. N., Yılmaz, K., & Doğru, H. (2020). Digitalization of business functions under industry 4.0. In *Digital Business Strategies in Blockchain Ecosystems* (pp. 105–132). Springer. doi:10.1007/978-3-030-29739-8_6

Calleja, D. (2019). Why the "New Plastics Economy" must be a circular economy. Field Actions Science Reports. *The Journal of Field Actions*, (19), 22-27.

Camero, A., & Alba, E. (2019). Smart City and information technology: A review. *Cities, 93*, 84-94.

Caniato, F., Caridi, M., Crippa, L., & Moretto, A. (2012). Environmental sustainability in fashion supply chains: An exploratory case based research. *International Journal of Production Economics*, *135*(2), 659–670. doi:10.1016/j.ijpe.2011.06.001

Caragliu, A., & Del Bo, C. F. (2019). Smart innovative cities: The impact of Smart City policies on urban innovation. *Technological Forecasting and Social Change*, *142*, 373–383. doi:10.1016/j.techfore.2018.07.022

Cesaretto, R., Buratto, A., & De Giovanni, P. (2021). Mitigating the feature fatigue effect for smart products through digital servitization. *Computers & Industrial Engineering*, *156*, 107218. doi:10.1016/j.cie.2021.107218

Chan, T. Y., & Wong, C. W. (2012). The consumption side of sustainable fashion supply chain: Understanding fashion consumer eco-fashion consumption decision. *Journal of Fashion Marketing and Management*, *16*(2), 193–215. doi:10.1108/13612021211222824

Circular, X. (2020). *Case study: Fairphone - Modular, long-life smartphones*. https://www.circularx.eu/en/cases/36/fairphone-modular-long-life-smartphones

Coates, G. W., & Getzler, Y. D. (2020). Chemical recycling to monomer for an ideal, circular polymer economy. *Nature Reviews. Materials*, *5*(7), 501–516. doi:10.103841578-020-0190-4

Daian, G., & Ozarska, B. (2009). Wood waste management practices and strategies to increase sustainability standards in the Australian wooden furniture manufacturing sector. *Journal of Cleaner Production*, *17*(17), 1594–1602. doi:10.1016/j.jclepro.2009.07.008

Damar, M. (2021). Metaverse Shape of Your Life for Future: A bibliometric snapshot. *Journal of Metaverse*, *1*(1), 1–8.

Dameri, R. P. (2014). Comparing smart and digital city: initiatives and strategies in Amsterdam and Genoa. Are they digital and/or smart? In *Smart city* (pp. 45–88). Springer.

de Boissieu, E., Kondrateva, G., Baudier, P., & Ammi, C. (2021). The use of blockchain in the luxury industry: Supply chains and the traceability of goods. *Journal of Enterprise Information Management*, *34*(5), 1318–1338. doi:10.1108/JEIM-11-2020-0471

De Brito, M. P., Carbone, V., & Blanquart, C. M. (2008). Towards a sustainable fashion retail supply chain in Europe: Organisation and performance. *International Journal of Production Economics*, *114*(2), 534–553. doi:10.1016/j.ijpe.2007.06.012

De Giovanni, P., & Folgiero, P. (2022). Strategies for the Circular Economy. Circular Districts and Networks. Giappichelli-Routledge. ISBN: 9788892123876.

De Giovanni, P. (2016). Coordination in a distribution channel with decisions on the nature of incentives and share-dependency on pricing. *The Journal of the Operational Research Society*, *67*(8), 1034–1049. doi:10.1057/jors.2015.118

De Giovanni, P. (2019). When feature-based production capabilities challenge operations: Drivers, moderators, and performance. *International Journal of Operations & Production Management*, *40*(2), 221–242. doi:10.1108/IJOPM-04-2019-0309

De GiovanniP. (2019a). *Eco-digital supply chains through blockchains*. Available at SSRN 3488925.

De Giovanni, P. (2020). An optimal control model with defective products and goodwill damages. *Annals of Operations Research*, *289*(2), 419–430. doi:10.100710479-019-03176-4

De Giovanni, P. (2020). *Dynamic quality models and games in digital supply chains*. Springer Nature.

De Giovanni, P. (2020a). Blockchain and smart contracts in supply chain management: A game theoretic model. *International Journal of Production Economics, 228*(10), 1–18. doi:10.1016/j.ijpe.2020.107855

De Giovanni, P. (2021). Smart contracts and blockchain for supply chain quality management. In *Dynamic Quality Models and Games in Digital Supply Chains* (pp. 91–110). Springer. doi:10.1007/978-3-030-66537-1_5

De Giovanni, P. (2022). Leveraging the circular economy with a closed-loop supply chain and a reverse omnichannel using blockchain technology and incentives. *International Journal of Operations & Production Management, 42*(7), 959–994. doi:10.1108/IJOPM-07-2021-0445

De Giovanni, P., & Folgiero, P. (2022). Circular Economy Strategies and the Emerging Trend of Circular Districts. *Routledge-Giapichelli.*

De Giovanni, P., & Folgiero, P. (2022). *Strategies for the Circular Economy*. Circular Districts and Networks. Giappichelli-Routledge.

De Giovanni, P., & Ramani, V. (2017). Product cannibalization and the effect of a service strategy. *The Journal of the Operational Research Society*, 1–17.

De Giovanni, P., & Vinzi, V. E. (2014). The benefits of a monitoring strategy for firms subject to the Emissions Trading System. *Transportation Research Part D, Transport and Environment, 33*, 220–233. doi:10.1016/j.trd.2014.06.008

De Giovanni, P., & Zaccour, G. (2022). A selective survey of game-theoretic models of closed-loop supply chains. *Annals of Operations Research, 314*(1), 77–116. doi:10.100710479-021-04483-5

Dedrick, J., & Kraemer, K. L. (2017). *Intangible assets and value capture in global value chains: The smartphone industry* (Vol. 41). WIPO.

Del Giudice, M., Chierici, R., Mazzucchelli, A., & Fiano, F. (2020). Supply chain management in the era of circular economy: The moderating effect of big data. *International Journal of Logistics Management, 32*(2), 337–356. doi:10.1108/IJLM-03-2020-0119

Delft, C. E. (2020). *Exploration chemical recycling–Extended summary. What is the potential contribution of chemical recycling to Dutch climate policy*. Academic Press.

di Bella, E., Odone, F., Corsi, M., Sillitti, A., & Breu, R. (2014). Smart Security: Integrated systems for security policies in urban environments. In *Smart city* (pp. 193–219). Springer.

Duan, H., Li, J., Fan, S., Lin, Z., Wu, X., & Cai, W. (2021, October). Metaverse for social good: A university campus prototype. In *Proceedings of the 29th ACM International Conference on Multimedia* (pp. 153-161). 10.1145/3474085.3479238

Duhaylongsod, J. B., & De Giovanni, P. (2018). The impact of innovation strategies on the relationship between supplier integration and operational performance. *International Journal of Physical Distribution & Logistics Management, 49*(2), 156–177. doi:10.1108/IJPDLM-09-2017-0269

Du, R., Santi, P., Xiao, M., Vasilakos, A. V., & Fischione, C. (2018). The sensable city: A survey on the deployment and management for smart city monitoring. *IEEE Communications Surveys and Tutorials, 21*(2), 1533–1560. doi:10.1109/COMST.2018.2881008

EchoChain. (2022). *The data behind Patagonia's no-logo decisions (and why it might be wrong).* Available at: www.echochain.com

European Environmental Agency. (2021). *Plastics, the circular economy and Europe¢s environment — A priority for action.* Publications Office of the European Union.

Fairphone. (2016). *Closing the Loop: The garbage collectors of the mobile industry.* https://www.fairphone.com/en/2016/07/14/closing-the-loop-the-garbage-collectors-of-the-mobile-industry/

Fairphone. (2018a). *Researching the Best Available Methods for Electronics Recycling.* Available at: https://www.fairphone.com/en/project/researching-best-available-methods-electronics-recycling/

Fairphone. (2018b). *Reuse and Recycling.* Available at: https://www.fairphone.com/en/impact/reuse-recycle/

Fairphone. (2021a). *Impact report - A challenge to the industry.* Available at: https://www.fairphone.com/wp-content/uploads/2021/06/Fairphone_Impact-report_2020.pdf

Fairphone. (2021b). *Impact Report 2020: Fairphone nearly doubles sales in 2020 as its industry influence increases.* Available at: https://www.fairphone.com/wp-content/uploads/2021/06/Impact-Report-2020-Fairphone-nearly-doubles-sales-in-2020-as-its-industry-influence-increases.pdf

Fairphone. (2021c). *Fair Materials.* Available at: https://www.fairphone.com/en/impact/fair-materials/

Falcone, P. (2018). *Fairphone, un'azienda ed uno smartphone stile libero: ecologico, equo, solidale e sostenibile.* Io Comunico. Available at: https://madeingaia.it/it/prodotti-sostenibili-persone/per-le-persone/comunichiamo/25-fairphone-un-azienda-ed-uno-smartphone-stile-libero-ecologico-equo-solidale-e-sostenibile

Farooque, M., Zhang, A., Thürer, M., Qu, T., & Huisingh, D. (2019). Circular supply chain management: A definition and structured literature review. *Journal of Cleaner Production, 228*, 882–900. doi:10.1016/j.jclepro.2019.04.303

FashionUnited. (2016). *Statistiche della moda Italia.* https://fashionunited.it/statistics/statistiche-moda-italia/

Feng, H., Wang, X., Duan, Y., Zhang, J., & Zhang, X. (2020). Applying blockchain technology to improve agri-food traceability: A review of development methods, benefits and challenges. *Journal of Cleaner Production, 260*(7), 1–15. doi:10.1016/j.jclepro.2020.121031

Fernqvist, F., Olsson, A., & Spendrup, S. (2015). What's in it for me? Food packaging and consumer responses, a focus group study. *British Food Journal, 117*(3), 1122–1135. doi:10.1108/BFJ-08-2013-0224

Fontana, F. (2014). The smart city and the creation of local public value. In *Smart City* (pp. 117–137). Springer.

Francisco, K., & Swanson, D. (2018). The supply chain has no clothes: Technology adoption of blockchain for supply chain transparency. *Logistics*, *2*(1), 2–8. doi:10.3390/logistics2010002

Furn. (2022). *Circular Economy in the furniture industry - Project 2017-1-BE01-KA202-024752*. Available at: https://www.furn360.eu/wp-content/uploads/2018/10/Circular-economy-in-the-furniture-industry-11092018.pdf

Fussler, C., & James, P. (1996). *Driving eco-innovation: a breakthrough discipline for innovation and sustainability*. Financial Times/Prentice Hall.

Galvez, J. F., Mejuto, J. C., & Simal-Gandara, J. (2018). Future challenges on the use of blockchain for food traceability analysis. *Trends in Analytical Chemistry*, *107*(10), 222–232. doi:10.1016/j.trac.2018.08.011

Gear, M., Sadhukhan, J., Thorpe, R., Clift, R., Seville, J., & Keast, M. (2018). A life cycle assessment data analysis toolkit for the design of novel processes–A case study for a thermal cracking process for mixed plastic waste. *Journal of Cleaner Production*, *180*, 735–747. doi:10.1016/j.jclepro.2018.01.015

Genovese, A., Acquaye, A. A., Figueroa, A., & Koh, S. L. (2017). Sustainable supply chain management and the transition towards a circular economy: Evidence and some applications. *Omega*, *66*, 344–357. doi:10.1016/j.omega.2015.05.015

Getor, R. Y., Mishra, N., & Ramudhin, A. (2020). The role of technological innovation in plastic production within a circular economy framework. *Resources, Conservation and Recycling*, *163*, 105094. doi:10.1016/j.resconrec.2020.105094

Geueke, B., Groh, K., & Muncke, J. (2018). Food packaging in the circular economy: Overview of chemical safety aspects for commonly used materials. *Journal of Cleaner Production*, *193*, 491–505. doi:10.1016/j.jclepro.2018.05.005

Ghisellini, P., Cialani, C., & Ulgiati, S. (2016). A review on circular economy: The expected transition to a balanced interplay of environmental and economic systems. *Journal of Cleaner Production*, *114*, 11–32. doi:10.1016/j.jclepro.2015.09.007

Giffinger, R., Fertner, C., Kramar, H., Kalasek, R., Pichler-Milanovic, N., & Meijers, E. (2007). *Smart cities. Ranking of European medium-sized cities*. Final Report, Centre of Regional Science, Vienna UT.

Global Recycling. (2021). https://global-recycling.info/archives/5931

Gonczol, P., Katsikouli, P., Herskind, L., & Dragoni, N. (2020). Blockchain implementations and use cases for supply chains-a survey. *IEEE Access: Practical Innovations, Open Solutions*, *8*(1), 11856–11871. doi:10.1109/ACCESS.2020.2964880

Grabowski, A., Selke, S. E., Auras, R., Patel, M. K., & Narayan, R. (2015). Life cycle inventory data quality issues for bioplastics feedstocks. *The International Journal of Life Cycle Assessment*, *20*(5), 584–596. doi:10.100711367-015-0853-3

Green Italy Report. (2021). *Un'economia a misura d'uomo per il futuro dell'Europa, Oct. 20th, 2021*. Author.

Green, S. G., & Taber, T. D. (1980). The effects of three social decision schemes on decision group process. *Organizational Behavior and Human Performance*, *25*(1), 97–106. doi:10.1016/0030-5073(80)90027-6

Greyson, J. (2007). An economic instrument for zero waste, economic growth and sustainability. *Journal of Cleaner Production*, *15*(13-14), 1382–1390. doi:10.1016/j.jclepro.2006.07.019

Gu, F., Hall, P., Miles, N. J., Ding, Q., & Wu, T. (2014). Improvement of mechanical properties of recycled plastic blends via optimizing processing parameters using the Taguchi method and principal component analysis. *Materials & Design, 62*, 189-198.

Gu, F., Guo, J., Zhang, W., Summers, P. A., & Hall, P. (2017). From waste plastics to industrial raw materials: A life cycle assessment of mechanical plastic recycling practice based on a real-world case study. *The Science of the Total Environment*, *601*, 1192–1207. doi:10.1016/j.scitotenv.2017.05.278 PMID:28605837

Gu, F., Hall, P., & Miles, N. J. (2016). Development of composites based on recycled polypropylene for injection moulding automobile parts using hierarchical clustering analysis and principal component estimate. *Journal of Cleaner Production*, *137*, 632–643. doi:10.1016/j.jclepro.2016.07.028

Gu, F., Hall, P., & Miles, N. J. (2016). Performance evaluation for composites based on recycled polypropylene using principal component analysis and cluster analysis. *Journal of Cleaner Production*, *115*, 343–353. doi:10.1016/j.jclepro.2015.12.062

Guide, V. D. R. Jr, & Van Wassenhove, L. N. (2006). Closed-loop supply chains: An introduction to the feature issue (part 1). *Production and Operations Management*, *15*(3), 345–350. doi:10.1111/j.1937-5956.2006.tb00249.x

Hader, M., Tchoffa, D., El Mhamedi, A., Ghodous, P., Dolgui, A., & Abouabdellah, A. (2022). Applying integrated Blockchain and Big Data technologies to improve supply chain traceability and information sharing in the textile sector. *Journal of Industrial Information Integration*, *28*(7), 1–20. doi:10.1016/j.jii.2022.100345

Harrison, C., Eckman, B., Hamilton, R., Hartswick, P., Kalagnanam, J., Paraszczak, J., & Williams, P. (2010). Foundations for smarter cities. *IBM Journal of Research and Development*, *54*(4), 1–16. doi:10.1147/JRD.2010.2048257

Hewa, T., Ylianttila, M., & Liyanage, M. (2021). Survey on blockchain based smart contracts: Applications, opportunities and challenges. *Journal of Network and Computer Applications*, *177*(3), 1–39. doi:10.1016/j.jnca.2020.102857

Hofmann, F., & Jaeger-Erben, M. (2020). Organizational transition management of circular business model innovations. *Business Strategy and the Environment, 29*(6), 2770–2788. doi:10.1002/bse.2542

Hottle, T. A., Bilec, M. M., & Landis, A. E. (2017). Biopolymer production and end of life comparisons using life cycle assessment. *Resources, Conservation and Recycling, 122*, 295–306. doi:10.1016/j.resconrec.2017.03.002

Hultberg, E., & Pal, R. (2021). Lessons on business model scalability for circular economy in the fashion retail value chain: Towards a conceptual model. *Sustainable Production and Consumption, 28*, 686–698. doi:10.1016/j.spc.2021.06.033

Ivanova, I. (2019). *Buying "green" is too pricey for the average consumer*. https://www.cbsnews.com/news/buying-green-is-too-pricey-for-the-average-consumer/

Jalali, H., Ansaripoor, A. H., & De Giovanni, P. (2020). Closed-loop supply chains with complementary products. *International Journal of Production Economics, 229*, 107757. doi:10.1016/j.ijpe.2020.107757

Jalali, H., Ansaripoor, A., Ramani, V., & De Giovanni, P. (2022). Closed-loop supply chain models with coopetition options. *International Journal of Production Research, 60*(10), 3078–3106. doi:10.1080/00207543.2021.1910871

Javed, T., Yang, J., Gilal, W. G., & Gilal, N. G. (2020). The sustainability claims' impact on the consumer's green perception and behavioral intention: A case study of H&M. *Advances in Management and Applied Economics, 10*(2), 1–22.

Jeswani, H., Krüger, C., Russ, M., Horlacher, M., Antony, F., Hann, S., & Azapagic, A. (2021). Life cycle environmental impacts of chemical recycling via pyrolysis of mixed plastic waste in comparison with mechanical recycling and energy recovery. *The Science of the Total Environment, 769*, 144483. doi:10.1016/j.scitotenv.2020.144483 PMID:33486181

Johansen, M. R., Christensen, T. B., Ramos, T. M., & Syberg, K. (2022). A review of the plastic value chain from a circular economy perspective. *Journal of Environmental Management, 302*, 113975. doi:10.1016/j.jenvman.2021.113975 PMID:34700081

Kagermann, H., Lukas, W. D., & Wahlster, W. (2011). Industrie 4.0: Mit dem Internet der Dinge auf dem Weg zur 4. industriellen Revolution. *VDI nachrichten, 13*(1), 2-3.

Kakadellis, S., & Harris, Z. M. (2020). Don't scrap the waste: The need for broader system boundaries in bioplastic food packaging life-cycle assessment–A critical review. *Journal of Cleaner Production, 274*, 122831. doi:10.1016/j.jclepro.2020.122831

Kan, M., & Miller, S. A. (2022). Environmental impacts of plastic packaging of food products. *Resources, Conservation and Recycling, 180*, 106156. doi:10.1016/j.resconrec.2022.106156

Kenniskaarten. (2013). *How is a circular economy different from a linear economy?* https://kenniskaarten.hetgroenebrein.nl/en/knowledge-map-circular-economy/how-is-a-circular-economy-different-from-a-linear-economy/

Ki, C. W., Chong, S. M., & Ha-Brookshire, J. E. (2020). How fashion can achieve sustainable development through a circular economy and stakeholder engagement: A systematic literature review. *Corporate Social Responsibility and Environmental Management*, *27*(6), 2401–2424. doi:10.1002/csr.1970

Kim, T. H., Ramos, C., & Mohammed, S. (2017). Smart city and IoT. *Future Generation Computer Systems*, *76*, 159–162. doi:10.1016/j.future.2017.03.034

Kitchin, R. (2018). The realtimeness of smart cities. *TECNOSCIENZA. Italian Journal of Science & Technology Studies*, *8*(2), 19–42.

Köhler, J., Sönnichsen, S. D., & Beske-Jansen, P. (2022). Towards a collaboration framework for circular economy: The role of dynamic capabilities and open innovation. *Business Strategy and the Environment*, *2*, 1–30. doi:10.1002/bse.3000

Köksal, D., Strähle, J., Müller, M., & Freise, M. (2017). Social sustainable supply chain management in the textile and apparel industry—A literature review. *Sustainability*, *9*(1), 100. doi:10.3390u9010100

Kortmann, S., & Piller, F. (2016). Open business models and closed-loop value chains: Redefining the firm-consumer relationship. *California Management Review*, *58*(3), 88–108. doi:10.1525/cmr.2016.58.3.88

Kouhizadeh, M., Zhu, Q., & Sarkis, J. (2020). Blockchain and the circular economy: Potential tensions and critical reflections from practice. *Production Planning and Control*, *31*(11-12), 950–966. doi:10.1080/09537287.2019.1695925

Kozderka, M., Rose, B., Kočí, V., Caillaud, E., & Bahlouli, N. (2016). High impact polypropylene (HIPP) recycling–Mechanical resistance and Lifecycle Assessment (LCA) case study with improved efficiency by preliminary sensitivity analysis. *Journal of Cleaner Production*, *137*, 1004–1017. doi:10.1016/j.jclepro.2016.07.217

Kshetri, N. (2018). 1 Blockchain's roles in meeting key supply chain management objectives. *International Journal of Information Management*, *39*(4), 80–89. doi:10.1016/j.ijinfomgt.2017.12.005

Kumar, V., & Dahiya, B. (2017). Smart economy in smart cities. In *Smart economy in smart cities* (pp. 3–76). Springer. doi:10.1007/978-981-10-1610-3_1

Kuo, T.-C., Huang, S. H., & Zhang, H.-C. (2001). Design for manufacture and design for 'X': Concepts, application and perspectives. *Computers & Industrial Engineering*, *41*(3), 241–260. doi:10.1016/S0360-8352(01)00045-6

Laufs, J., Borrion, H., & Bradford, B. (2020). Security and the smart city: A systematic review. *Sustainable Cities and Society*, *55*, 102023. doi:10.1016/j.scs.2020.102023

Lazarevic, D., Aoustin, E., Buclet, N., & Brandt, N. (2010). Plastic waste management in the context of a European recycling society: Comparing results and uncertainties in a life cycle perspective. *Resources, Conservation and Recycling*, *55*(2), 246–259. doi:10.1016/j.resconrec.2010.09.014

Lee, J., Azamfar, M., & Singh, J. (2019). A blockchain enabled Cyber-Physical System architecture for Industry 4.0 manufacturing systems. *Manufacturing Letters, 20*(4), 34–39. doi:10.1016/j.mfglet.2019.05.003

Lee, L. H., Braud, T., Zhou, P., Wang, L., Xu, D., Lin, Z., ... Hui, P. (2021). All one needs to know about metaverse: A complete survey on technological singularity, virtual ecosystem, and research agenda. *Journal of Latex Class Files, 14*(8).

Leng, J., Ruan, G., Jiang, P., Xu, K., Liu, Q., Zhou, X., & Liu, C. (2020). Blockchain-empowered sustainable manufacturing and product lifecycle management in industry 4.0: A survey. *Renewable & Sustainable Energy Reviews, 132*(10), 1–22. doi:10.1016/j.rser.2020.110112

Levi Strauss & Co. (2014). https://www.levistrauss.com/2014/03/13/denim-insulation-101-how-recycled-jeans-can-keep-you-warm/

Levi Strauss & Co. (2018). https://www.levistrauss.com/wp-content/uploads/2018/07/LSCO_Climate_Action_Strategy_2025.pdf

Levi Strauss & Co. (2020a). https://www.levistrauss.com/sustainability-report/our-reporting/our-stakeholders/

Levi Strauss & Co. (2020b). https://www.levistrauss.com/wp-content/uploads/2021/09/2020-LSCo.-GRI-SDG-UNGC-Index.pdf

Levi Strauss & Co. (2020c). https://www.levistrauss.com/wp-content/uploads/2020/07/LSCo.-2019-Sustainability-Review.pdf

Levi Strauss & Co. (2021a). https://www.levistrauss.com/2021/07/21/designing-a-more-circular-future/

Levi Strauss & Co. (2021b). https://www.levistrauss.com/sustainability-report/our-reporting/report-in-brief/

Levi Strauss & Co. (2021c). https://www.levistrauss.com/wp-content/uploads/2021/09/2020-LSCo.-SASB-Table.pdf

Lieder, M., & Rashid, A. (2016). Towards circular economy implementation: A comprehensive review in context of manufacturing industry. *Journal of Cleaner Production, 115*, 36–51. doi:10.1016/j.jclepro.2015.12.042

Liu, B., & De Giovanni, P. (2019). Green process innovation through Industry 4.0 technologies and supply chain coordination. *Annals of Operations Research*, 1–36. doi:10.100710479-019-03498-3

MacArthur, E. (2013). Towards the circular economy. *Journal of Industrial Ecology, 2*(1), 23–44.

MacArthur, E., Zumwinkel, K., & Stuchtey, M. R. (2015). *Growth within: A circular economy vision for a competitive Europe*. Ellen MacArthur Foundation.

Maddah, H. A. (2016). Polypropylene as a promising plastic: A review. *American Journal of Political Science, 6*(1), 1–11.

Mango, R. (2020). *Sustainable practices in plastic waste management – case study of Acqua San Benedetto SpA*. Masarykova univerzita, Ekonomicko-správní fakulta; Masarykova univerzita; Masaryk University. Retrieved from https://theses.cz/id/7f782u/

Maranesi, C., & De Giovanni, P. (2020). Modern circular economy: Corporate strategy, supply chain, and industrial symbiosis. *Sustainability*, *12*(22), 9383. doi:10.3390u12229383

Matthews, C., Moran, F., & Jaiswal, A. K. (2021). A review on European Union's strategy for plastics in a circular economy and its impact on food safety. *Journal of Cleaner Production*, *283*, 125263. doi:10.1016/j.jclepro.2020.125263

Meijer, A., & Bolívar, M. P. R. (2016). Governing the smart city: A review of the literature on smart urban governance. *International Review of Administrative Sciences*, *82*(2), 392–408. doi:10.1177/0020852314564308

Mendes, A. C., & Pedersen, G. A. (2021). Perspectives on sustainable food packaging:–is bio-based plastics a solution? *Trends in Food Science & Technology*, *112*, 839–846. doi:10.1016/j.tifs.2021.03.049

Moss Kanter, R., & Litow, S. S. (2009). *Informed and interconnected: A manifesto for smarter cities*. Harvard Business School General Management Unit Working Paper (09-141).

Murray, A., Skene, K., & Haynes, K. (2017). The circular economy: An interdisciplinary exploration of the concept and application in a global context. *Journal of Business Ethics*, *140*(3), 369–380. doi:10.100710551-015-2693-2

Musti, K. S. (2020). Circular economy in energizing smart cities. In *Handbook of Research on Entrepreneurship Development and Opportunities in Circular Economy* (pp. 251–269). IGI Global. doi:10.4018/978-1-7998-5116-5.ch013

Naclerio, A. G., & De Giovanni, P. (2022). Blockchain, logistics and omnichannel for last mile and performance. *International Journal of Logistics Management*, *33*(2), 663–686. doi:10.1108/IJLM-08-2021-0415

Nessi, S., Bulgheroni, C., Konti, A., Sinkko, T., Tonini, D., & Pant, R. (2018). *Environmental sustainability assessment comparing through the means of lifecycle assessment the potential environmental impacts of the use of alternative feedstock (biomass, recycled plastics, CO2) for plastic articles in comparison to using current feedstock (oil and gas)*. Draft Report for Stakeholder Consultation (Part 2).

Nevelsteen, K. J. (2018). Virtual world, defined from a technological perspective and applied to video games, mixed reality, and the Metaverse. *Computer Animation and Virtual Worlds*, *29*(1), e1752. doi:10.1002/cav.1752

Nielsen, T. D., Hasselbalch, J., Holmberg, K., & Stripple, J. (2020). Politics and the plastic crisis: A review throughout the plastic life cycle. *Wiley Interdisciplinary Reviews. Energy and Environment*, *9*(1), e360. doi:10.1002/wene.360

Noori, N., de Jong, M., Janssen, M., Schraven, D., & Hoppe, T. (2021). Input-output modeling for smart city development. *Journal of Urban Technology*, *28*(1-2), 71–92. doi:10.1080/10630732.2020.1794728

Nussholz, J. L. (2018). A circular business model mapping tool for creating value from prolonged product lifetime and closed material loops. *Journal of Cleaner Production*, *197*, 185–194. doi:10.1016/j.jclepro.2018.06.112

Ortega-Fernández, A., Martín-Rojas, R., & García-Morales, V. J. (2020). Artificial intelligence in the urban environment: Smart cities as models for developing innovation and sustainability. *Sustainability*, *12*(19), 7860. doi:10.3390u12197860

Otto, S., Strenger, M., Maier-Nöth, A., & Schmid, M. (2021). Food packaging and sustainability–Consumer perception vs. correlated scientific facts: A review. *Journal of Cleaner Production*, *298*, 126733. doi:10.1016/j.jclepro.2021.126733

Pal, R. (2017). Value creation through reverse logistics in used clothing networks. *International Journal of Logistics Management*, *28*(3), 864–906. doi:10.1108/IJLM-11-2016-0272

Perrault, M. (2021). *How Companies Can Overcome the Barriers to Sustainability in the Fashion Industry*. https://scholarworks.uark.edu/cgi/viewcontent.cgi?article=1057&context=finnuht

Petit-Boix, A., & Leipold, S. (2018). Circular economy in cities: Reviewing how environmental research aligns with local practices. *Journal of Cleaner Production*, *195*, 1270–1281. doi:10.1016/j.jclepro.2018.05.281

PETnology. (2021). *Acqua Minerale San Benedetto launches the new Ecogreen 1l Easy Bottle, made with 100% recycled PET*. Retrieved from https://www.petnology.com/competence-online/news/acqua-minerale-san-benedetto-launches-the-new-ecogreen-1l-easy-bottle-made-with-100-recycled-pet.html

Phelan, A. A., Meissner, K., Humphrey, J., & Ross, H. (2022). Plastic pollution and packaging: Corporate commitments and actions from the food and beverage sector. *Journal of Cleaner Production*, *331*, 129827. doi:10.1016/j.jclepro.2021.129827

Placido Volpone. (2019). https://placidovolpone.it/blockchain-vini/

Preece. (1995). *Building a Reputation Along the Value Chain at Levi Strauss*. Academic Press.

Prencipe, M. P., Vishkaei, B. M., & De Giovanni, P. (2022). Blockchain Adoption in the Winery Industry: The Case of Cantina Placido-Volpone. In Blockchain Technology Applications in Businesses and Organizations (pp. 43-72). IGI Global.

Ragaert, K., Delva, L., & Van Geem, K. (2017). Mechanical and chemical recycling of solid plastic waste. *Waste Management, 69*, 24-58.

Rahim, R., & Raman, A. A. A. (2017). Carbon dioxide emission reduction through cleaner production strategies in a recycled plastic resins producing plant. *Journal of Cleaner Production*, *141*, 1067–1073. doi:10.1016/j.jclepro.2016.09.023

Rajendran, S., Hodzic, A., Scelsi, L., Hayes, S., Soutis, C., Alma'Adeed, M., & Kahraman, R. (2013). Plastics recycling: Insights into life cycle impact assessment methods. *Plastics, Rubber and Composites*, *42*(1), 1–10. doi:10.1179/1743289812Y.0000000002

Ramani, V., & De Giovanni, P. (2017). A two-period model of product cannibalization in an atypical Closed-loop Supply Chain with endogenous returns: The case of DellReconnect. *European Journal of Operational Research*, *262*(3), 1009–1027. doi:10.1016/j.ejor.2017.03.080

Rejeb, A., Keogh, J. G., Zailani, S., Treiblmaier, H., & Rejeb, K. (2020). Blockchain technology in the food industry: A review of potentials, challenges and future research directions. *Logistics*, *4*(4), 1–26. doi:10.3390/logistics4040027

Renewcell. (2020a). https://www.renewcell.com/en/circulose/

Renewcell. (2020b). https://www.renewcell.com/en/section/about-renewcell/

Reuter, M. A., van Schaik, A., & Ballester, M. (2018). Limits of the circular economy: Fairphone modular design pushing the limits. *World Metall.–. Erzmetall*, *71*(2), 68–79.

Rinnovabili e Risparmio. (2019). *Sostenibilità l'impatto ambientale del fast fashion.* https://www.rinnovabilierisparmio.it/2019/01/11/sostenibilita-l-impatto-ambientale-della-fast-fashion/

Risius, M., & Spohrer, K. (2017). A blockchain research framework. *Business & Information Systems Engineering*, *59*(6), 385–409. doi:10.100712599-017-0506-0

Rundh, B. (2016). The role of packaging within marketing and value creation. *British Food Journal*, *118*(10), 2491–2511. doi:10.1108/BFJ-10-2015-0390

Saberi, S., Kouhizadeh, M., & Sarkis, J. (2019). Blockchains and the supply chain: Findings from a broad study of practitioners. *IEEE Engineering Management Review*, *47*(3), 95–103. doi:10.1109/EMR.2019.2928264

Sacco, A., & De Giovanni, P. (2019). Channel coordination with a manufacturer controlling the price and the effect of competition. *Journal of Business Research*, *96*, 97–114. doi:10.1016/j.jbusres.2018.09.001

Saeed, M. A., Waseek, I., & Kersten, W. (2017). Literature review of drivers of sustainable supply chain management. *Proceedings of the Hamburg international Conference of Logistics (HICL)*, 159-184.

Sandberg, E., Pal, R., & Hemilä, J. (2018). Exploring value creation and appropriation in the reverse clothing supply chain. *International Journal of Logistics Management*, *29*(1), 90–109. doi:10.1108/IJLM-10-2016-0241

Schmidt, C. G., Klöckner, M., & Wagner, S. M. (2021). Blockchain for supply chain traceability: case examples for luxury goods. In *Digital Business Models in Industrial Ecosystems* (pp. 187–197). Springer. doi:10.1007/978-3-030-82003-9_12

Seuring, S., & Müller, M. (2008). From a literature review to a conceptual framework for sustainable supply chain management. *Journal of Cleaner Production, 16*(15), 1699–1710. doi:10.1016/j.jclepro.2008.04.020

Shatri, A. (2021). *Complete SWOT Analysis of Patagonia. One of the most sustainable and eco-friendly retailers.* The Digital School. Available at: https://iide.co/case-studies/swot-analysis-of-patagonia/

Shen, B. (2014). Sustainable fashion supply chain: Lessons from H&M. *Sustainability, 6*(9), 6236–6249. doi:10.3390u6096236

Shen, B., Zheng, J. H., Chow, P. S., & Chow, K. Y. (2014). Perception of fashion sustainability in online community. *Journal of the Textile Institute, 105*(9), 971–979. doi:10.1080/00405000.2013.866334

Sheriff, K. M., Subramanian, N., Rahman, S., & Jayaram, J. (2017). Integrated optimization model and methodology for plastics recycling: Indian empirical evidence. *Journal of Cleaner Production, 153*, 707–717. doi:10.1016/j.jclepro.2016.07.137

Shi, S., Wang, L., Calvert, K. L., & Griffioen, J. N. (2004, April). A multi-path routing service for immersive environments. In *IEEE International Symposium on Cluster Computing and the Grid* (pp. 699-706). IEEE.

Singh, N., Hui, D., Singh, R., Ahuja, I. P. S., Feo, L., & Fraternali, F. (2017). Recycling of plastic solid waste: A state of art review and future applications. *Composites. Part B, Engineering, 115*, 409–422. doi:10.1016/j.compositesb.2016.09.013

Social Share. (2021). https://www.commonobjective.co/article/the-issues-water

Sommerhuber, P. F., Wang, T., & Krause, A. (2016). Wood–plastic composites as potential applications of recycled plastics of electronic waste and recycled particleboard. *Journal of Cleaner Production, 121*, 176–185. doi:10.1016/j.jclepro.2016.02.036

Somoza-Tornos, A., Gonzalez-Garay, A., Pozo, C., Graells, M., Espuña, A., & Guillén-Gosálbez, G. (2020). Realizing the potential high benefits of circular economy in the chemical industry: Ethylene monomer recovery via polyethylene pyrolysis. *ACS Sustainable Chemistry & Engineering, 8*(9), 3561–3572. doi:10.1021/acssuschemeng.9b04835

Srivastava, S. K. (2007). Green supply-chain management: A state-of-the-art literature review. *International Journal of Management Reviews, 9*(1), 53–80. doi:10.1111/j.1468-2370.2007.00202.x

Stål, H. I., & Corvellec, H. (2018). A decoupling perspective on circular business model implementation: Illustrations from Swedish apparel. *Journal of Cleaner Production, 171*, 630–643. doi:10.1016/j.jclepro.2017.09.249

Steenis, N. D., Van Herpen, E., Van Der Lans, I. A., Ligthart, T. N., & Van Trijp, H. C. (2017). Consumer response to packaging design: The role of packaging materials and graphics in sustainability perceptions and product evaluations. *Journal of Cleaner Production, 162*, 286–298. doi:10.1016/j.jclepro.2017.06.036

Stephenson, N. (1992). *Snow crash: A novel*. Bantam Books.

Stichnothe, H., & Azapagic, A. (2013). Life cycle assessment of recycling PVC window frames. *Resources, Conservation and Recycling*, *71*, 40–47. doi:10.1016/j.resconrec.2012.12.005

Suárez-Eiroa, B., Fernández, E., Méndez-Martínez, G., & Soto-Oñate, D. (2019). Operational principles of circular economy for sustainable development: Linking theory and practice. *Journal of Cleaner Production*, *214*, 952–961. doi:10.1016/j.jclepro.2018.12.271

The European House, Ambrosetti. (2013). *L'eccellenza della filiera della plastica per il rilancio industriale dell'Italia e dell'Europa*. Author.

Thomason, J. (2021). MetaHealth-How will the Metaverse Change Health Care? *Journal of Metaverse*, *1*(1), 13–16.

Trabucco, M., & De Giovanni, P. (2021). Achieving Resilience and Business Sustainability during COVID-19: The Role of Lean Supply Chain Practices and Digitalization. *Sustainability*, *13*(22), 12369. doi:10.3390u132212369

Tseng, M. L., Tan, R. R., Chiu, A. S., Chien, C. F., & Kuo, T. C. (2018). Circular economy meets industry 4.0: Can big data drive industrial symbiosis? *Resources, Conservation and Recycling*, *131*, 146–147. doi:10.1016/j.resconrec.2017.12.028

UNEP. (2020). *Plastic Waste – Background Report*. UNEP.

Upadhyay, A., Mukhuty, S., Kumar, V., & Kazancoglu, Y. (2021). Blockchain technology and the circular economy: Implications for sustainability and social responsibility. *Journal of Cleaner Production*, *293*(4), 1–7. doi:10.1016/j.jclepro.2021.126130

Van der Land, S., Schouten, A., & Feldberg, F. (2011). Modeling the metaverse: A theoretical model of effective team collaboration in 3D virtual environments. *Journal of Virtual Worlds Research*, *4*(3). Advance online publication. doi:10.4101/jvwr.v4i3.6126

Vernaza, A., Armuelles, V. I., & Ruiz, I. (2012, June). Towards to an open and interoperable virtual learning enviroment using Metaverse at University of Panama. In 2012 Technologies Applied to Electronics Teaching (TAEE) (pp. 320-325). IEEE. doi:10.1109/TAEE.2012.6235458

Vishkaei, B. M., Fathi, M., Khakifirooz, M., & De Giovanni, P. (2021). Bi-objective optimization for customers' satisfaction improvement in a Public Bicycle Sharing System. *Computers & Industrial Engineering*, *161*, 107587. doi:10.1016/j.cie.2021.107587

Vishkaei, B. M., Mahdavi, I., Mahdavi-Amiri, N., & Khorram, E. (2020). Balancing public bicycle sharing system using inventory critical levels in queuing network. *Computers & Industrial Engineering*, *141*, 106277. doi:10.1016/j.cie.2020.106277

Wäger, P. A., & Hischier, R. (2015). Life cycle assessment of post-consumer plastics production from waste electrical and electronic equipment (WEEE) treatment residues in a Central European plastics recycling plant. *The Science of the Total Environment*, *529*, 158–167. doi:10.1016/j.scitotenv.2015.05.043 PMID:26022405

Walker, S., & Rothman, R. (2020). Life cycle assessment of bio-based and fossil-based plastic: A review. *Journal of Cleaner Production, 261*, 121158. doi:10.1016/j.jclepro.2020.121158

Walter, G., Knizek, C., von Koeller, E., O'Brien, C., Hardin, E. M., Young, D., & Cordes, F. (2020). *Your supply chain needs a sustainability strategy*. Academic Press.

Wernink, T., & Strahl, C. (2015). Fairphone: Sustainability from the inside-out and outside-. In *Sustainable value chain management* (pp. 123–139). Springer. doi:10.1007/978-3-319-12142-0_3

White, A., & Lockyer, S. (2020). Removing plastic packaging from fresh produce–what's the impact? *Nutrition Bulletin, 45*(1), 35–50. doi:10.1111/nbu.12420

WoodCircus. (2021). *Holy Wood*. Available at: https://woodcircus.eu/wp-content/uploads/2021/01/04-Holy-wood-WoodCircus-p.pdf

WWF Italia Onlus. (2021). *Plastic Smart Cities*. Available at https://www.wwf.it/cosa-facciamo/progetti/plastic-smart-cities/

Yates, M. R., & Barlow, C. Y. (2013). Life cycle assessments of biodegradable, commercial biopolymers—A critical review. *Resources, Conservation and Recycling, 78*, 54–66. doi:10.1016/j.resconrec.2013.06.010

Yeh, H. (2017). The effects of successful ICT-based smart city services: From citizens' perspectives. *Government Information Quarterly, 34*(3), 556–565. doi:10.1016/j.giq.2017.05.001

Yigitcanlar, T., Kankanamge, N., & Vella, K. (2021). How are smart city concepts and technologies perceived and utilized? A systematic geo-Twitter analysis of smart cities in Australia. *Journal of Urban Technology, 28*(1-2), 135–154. doi:10.1080/10630732.2020.1753483

Yin, S., Tuladhar, R., Shi, F., Shanks, R. A., Combe, M., & Collister, T. (2015). Mechanical reprocessing of polyolefin waste: A review. *Polymer Engineering and Science, 55*(12), 2899–2909. doi:10.1002/pen.24182

Zero Panik. (2018). https://zeropanik.com/

Zhang, Y., & Zhou, P. (2021). Consumer Attitude towards Sustainability of Fast Fashion Products in the UK. *Sustainability, 13*(4), 1646. Advance online publication. doi:10.3390u13041646

Zhou, K., Liu, T., & Zhou, L. (2015). Industry 4.0: Towards future industrial opportunities and challenges. *2015 12th International conference on fuzzy systems and knowledge discovery (FSKD)*, 2147-2152.

Zsidisin, G. A., & Siferd, S. P. (2001). Environmental purchasing: A framework for theory development. *European Journal of Purchasing & Supply Management, 7*(1), 61–73. doi:10.1016/S0969-7012(00)00007-1

Zuo, Y. (2021). Making smart manufacturing smarter–a survey on blockchain technology in Industry 4.0. *Enterprise Information Systems, 15*(10), 1323–1353. doi:10.1080/17517575.2020.1856425

About the Contributors

Pietro De Giovanni is Professor of Operations Management, Circular Economy and Sustainable Strategies in the Department of Business and Management at Luiss University, and Associate Dean for Sustainability at Luiss-X.ITE Research Center. He holds a PhD from ESSEC Business School, and was visiting PhD candidate at GERAD, HEC Montréal. Before joining Luiss, he has held positions at ESSEC Business School (Paris-Singapore), Vrije Universiteit (Amsterdam, the Netherlands), and NOVA School of Business and Economics (Lisbon, Portugal). His research interests are circular economy, supply chain, operations management, sustainability, and digital transformation. He is Associate Editor of both the International Transactions of Operations Research and 4OR-Springer. He has published papers in Journal of Operations Management, European Journal of Operational Research, Transportation Research, Annals of Operations Research, International Journal of Production Economics, Journal of Business Research, Omega, and Research Policy.

* * *

Louis Alarcon is a double degree student in Management and Design (Master) at Grenoble École de Management and Strate École de Design (France).

Sofia Brunelli was born in Rome in 1998. She graduated in Political Science and International Relations at Luiss Guido Carli University where she is also attending the MSc in Law, digital innovation, and sustainability. She is passionate about geopolitics of climate change and more in general environmental topics.

Alessandra Cozzolino, PhD, is an Associate Professor of Management at Sapienza University of Rome, Faculty of Economics, Department of Management, where she teaches "Supply chain management", and "Business-to-business market management". Her main research fields are management, logistics & supply chain management, innovation. She is member of the Editorial Advisory Board of the Journal of Humanitarian Logistics and Supply Chain Management, of the Editorial Review Board of the Journal of Business Logistics, of the Scientific Committee of the book series "Themes and cases in logistics and supply chain management", of the Board of Sapienza Doctoral Program in "Management, Banking and Commodity Sciences", of Sapienza University Research Commission. She is President of the Monitoring Committee on Quality of the Faculty of Economics. She is author of many international and national scientific publications. She is principal investigator of several Sapienza University funded projects on logistics and supply chain management. She collaborates with Sapienza School for Advanced Studies. She has been Visiting research scholar in Supply chain management at Warrington College of Business Administration, University of Florida.

Riccardo De Lucia is a Luiss Guido Carli Student of Marketing Analytics and Metrics.

Chiara Fusari is a second-year student of the master in Law, Digital Innovation and Sustainability offered by the Luiss Guido Carli University. With a bachelor degree in economics, she is interested in understanding the new social and environmental trends in the economy, and how business administration is evolving accordingly.

Behzad Maleki Vishkaei is an Assistant Professor at the Department of Business and Management. Prior to joining Luiss, he was an Assistant Professor at the University of Eyvanekey (Iran), and before that he used to teach as an instructor at several universities in Iran. He cooperated also with different companies as a Strategic Planning and Project Management expert. He holds a Ph.D. in Industrial Engineering from Mazandaran University of Science and Technology (Iran). His research topics focus on Supply Chain Management, Operations Research, Transportation, Digital Transformation, and Smart Cities. He is interested in developing practical models and solutions that can be useful for companies and cities.

Francesca Nacca is a second year student of the Master in Law, Digital Innovation and Sustainability at Luiss Guido Carli University. With a bachelor's degree in Economics, she is now completing her Master's studies with a view to graduating in July 2022. Francesca is very interested in topics related to digital and energy transitions and their social, economic and environmental benefits.

Benedicta Naa Sackiorkor Quarcoo is a final year Master of Law, Digital Innovation & Sustainability student with a major in Sustainability at Luiss Guido Carli University in Rome, Italy. She hails from Africa, specifically from Ghana, and attained her Bachelor's degree from one of the prestigious universities in Ghana, the University of Ghana in Accra, Ghana. She seeks to become a Sustainability expert, aiding in any way possible to help achieve the targets of the UN Sustainable Development Goals to make the world a more livable and conducive environment for all living organisms.

Vinay Ramani is an Associate Professor at Indian Institute of Technology Kanpur. His research interests are in the Economics-Operations Management interface, particularly, game theoretic models of closed-loop supply chains, product cannibalization, and platform competition,

Nandan Kumar Singh is a Doctoral Candidate in the Production and Operations Management area at the Indian Institute of Management Visakhapatnam. His research interests are in closed-loop supply chains, consumer returns, Supply chain risk and resilience, Additive Manufacturing, and Circular Economy.

Index

www.ingramcontent.com/pod-product-compliance
Lightning Source LLC
LaVergne TN
LVHW061542290926
849162LV00004B/881
9781668450017